Conference Board of the Mathematical Sciences

CBMS

Regional Conference Series in Mathematics

Number 64

Group Rings, Crossed Products and Galois Theory

Donald S. Passman

Published for the
Conference Board of the Mathematical Sciences
by the
American Mathematical Society
Providence, Rhode Island
with support from the
National Science Foundation

Expository Lectures
from the CBMS Regional Conference
held at Mankato State University
June 17–21, 1985

The conference was supported in part by National Science Foundation Grant DMS-8503713.

1991 *Mathematics Subject Classification*. Primary 16S35;
Secondary 16S34, 16W20.

Library of Congress Cataloging-in-Publication Data
Passman, Donald S., 1940–
   Group rings, crossed products, and Galois theory.
     p. cm. — (Regional conference series in mathematics/Conference Board of the Mathematical Sciences, ISSN 0160-7642; no. 64)
   "Expository lectures from the CBMS regional conference held at Mankato State University, June 17–21, 1985"—T.p. verso.
   Conference sponsored by CBMS and the National Science Foundation.
   Includes Bibliographies.
   1. Associative rings. 2. Group rings. 3. Galois theory. I. Conference Board of the Mathematical Sciences. II. National Science Foundation (U.S.) III. Title. IV. Series: Regional conference series in mathematics; no. 64.
QA1.R33 no. 64   510s   [512'.4]   86-1177
[QA251.5]
ISBN 0-8218-0714-5

Copying and reprinting. Individual readers of this publication, and nonprofit libraries acting for them, are permitted to make fair use of the material, such as to copy a chapter for use in teaching or research. Permission is granted to quote brief passages from this publication in reviews, provided the customary acknowledgment of the source is given.

Republication, systematic copying, or multiple reproduction of any material in this publication (including abstracts) is permitted only under license from the American Mathematical Society. Requests for such permission should be addressed to the Assistant to the Publisher, American Mathematical Society, P.O. Box 6248, Providence, Rhode Island 02940-6248. Requests can also be made by e-mail to reprint-permission@ams.org.

© Copyright 1986 by the American Mathematical Society. All rights reserved.
The American Mathematical Society retains all rights
except those granted to the United States Government.
Printed in the United States of America.

∞ The paper used in this book is acid-free and falls within the guidelines
established to ensure permanence and durability.
♻ Printed on recycled paper.

10 9 8 7 6 5 4 3 2   01 00 99 98 97 96

Contents

Acknowledgments	v
Introduction	vii
1. Δ-Methods in Group Rings	1
2. The Jacobson Radical of Group Rings	7
3. Zero Divisors in Group Rings	14
4. Polycyclic Group Rings	22
5. Crossed Products of Finite Groups	29
6. Crossed Products of Infinite Groups	37
7. Computing the Symmetric Ring of Quotients	44
8. Galois Theory and Crossed Products	51
9. Galois Theory of Prime Rings	58
10. Rings and Fixed Rings	65

Acknowledgments

I would like to thank the CBMS and the NSF for sponsoring this conference, Frank Hannick for his fine management of it and Dan Farkas, Ed Formanek, Susan Montgomery, Lance Small and Robert Warfield for giving the supporting lectures.

Towards the end of this meeting I became quite ill. I wish to express my appreciation to all the participants of the conference, especially to Vic Camillo, Frank Hannick, Jeanne Kerr and Sylvia Wiegand, for their kindness and concern. I would also like to thank Susan Montgomery for giving my last four lectures when I was unable to do so.

Finally, love to my family, Marj, Barbara and Jon, for all things nonmathematical.

Introduction

During the week of June 17, 1985 a conference was held at Mankato State University in Mankato, Minnesota. This monograph contains slightly expanded versions of the ten talks I wrote for the occasion. Their theme is the interplay between group theory and ring theory. Specifically, they concern group rings, crossed products and the Galois theory of rings.

Group rings. The group ring $R[G]$ is an easily defined, attractive object to study. It is a tool of group theory, a tool of ring theory and an interesting subject in its own right. If G is a finite group and R is a field, then the subject concerns the ordinary and modular representations and characters of G. If R is the ring of integers, then we are dealing with the integral representations of G and its close relationship with maximal orders and algebraic number theory.

If G is polycyclic-by-finite and R is a Noetherian ring, then so is $R[G]$. Thus we obtain a large class of interesting examples of such rings, and indeed this subject is now closely allied with the general study of noncommutative Noetherian rings. Furthermore, there are many analogies here with enveloping algebras of finite-dimensional Lie algebras.

For G arbitrary, it is best to restrict our attention to R being a field K or the ring of integers. However, even with this assumption, we can no longer use finite techniques and the problems become correspondingly more difficult. The subject of group algebras of infinite groups was begun in the 50's and 60's with work of G. Higman, I. Kaplansky and S. A. Amitsur. It was most active in the 70's. The first four lectures are devoted to this material.

Crossed products. In some sense the crossed product $R * G$ is a generalized group ring, but it is actually much more. It first occurred in the study of division rings and, in particular, of the Brauer group of fields. A classical question with a modern answer due to Amitsur concerned whether all finite-dimensional division rings are crossed products. In another vein, if one is interested in the Galois theory of rings, then the skew group ring RG is an important tool. It of course contains all the ingredients of the theory, namely the ring R, the group G and the fixed ring R^G. Results on crossed products have had impressive Galois theoretic applications.

Furthermore, even if one is only interested in ordinary group algebras $K[G]$, then crossed products still occur. For example, if N is a normal subgroup of G, then $K[G] = K[N] * (G/N)$ is a suitable crossed product of the quotient group G/N over the subring $K[N]$. In addition, it frequently happens that if P is a prime ideal of $K[G]$, then $K[G]/P$ has a natural crossed product structure. For these and many other reasons, crossed products are interesting and important objects of study. Three lectures are devoted to this subject, emphasizing fairly recent work.

Galois theory. The Galois theory of noncommutative rings is a new subject with roots in invariant theory and the classical Galois theory of fields and division rings. If G acts on a ring R, then we are concerned with the relationship between R and the fixed ring R^G, and the relationship between the subgroups of G and the intermediate rings.

One can argue that the subject began in the early 70's with the important Bergman-Isaacs theorem on the existence of fixed points. This was later extended by V. K. Kharchenko who also developed a general Galois theory for semiprime rings. Important contributions were made by J. W. Fisher, S. Montgomery, J. Osterburg and many others. Actions on specific rings have also been studied. Of particular interest are actions on free rings, generic matrix rings, polynomial rings, enveloping algebras and group algebras. The entire subject is quite active at present and is being pursued in many different directions. The final three lectures consider selected aspects of this material.

The ten talks included here are essentially independent although the notation builds up. Each contains a selection of results on a particular subject, a limited number of proofs or sketches and at least a few open questions. Furthermore, each has its own brief list of references. Two general references for the series as a whole are S. Montgomery's monograph, *Fixed Rings of Finite Automorphism Groups of Associative Rings*, and my book, *The Algebraic Structure of Group Rings*. Finally, since the topics chosen are all part of a common theme, they do begin to merge and the lectures are eventually all interrelated.

1. Δ-Methods in Group Rings

We begin by discussing group algebras. Let K be a field and let G be a multiplicative group. Then the group algebra $K[G]$ is a K-vector space with basis G. Thus every element of $K[G]$ is a formal finite sum $\alpha = \sum_{x \in G} a_x x$ with each $a_x \in K$. Addition in $K[G]$ is of course componentwise, and multiplication is defined distributively using the multiplication of G. In this way, $K[G]$ becomes an associative K-algebra.

If α is as above, then its support is given by

$$\operatorname{Supp} \alpha = \{x \in G \mid a_x \neq 0\}.$$

Thus $\operatorname{Supp} \alpha$ is a finite subset of G. If S is any subset of G, then we let $K[S]$ denote the K-subspace of $K[G]$ spanned by the elements of S. Furthermore, there is a natural projection map $\pi_S : K[G] \to K[S]$ defined by

$$\pi_S \left(\sum_{x \in G} a_x x \right) = \sum_{x \in S} a_x x.$$

When $S = \{1\}$, we frequently write $\pi_S = \operatorname{tr}$ because $\operatorname{tr} : K[G] \to K$ enjoys many trace-like properties. When S is finite we let $\hat{S} = \sum_{s \in S} s$ denote the sum of the elements of S so that $\hat{S} \in K[S] \subseteq K[G]$. Observe that if $S = H$ is a subgroup of G, then $K[H]$ is the group algebra of H naturally embedded in $K[G]$.

LEMMA 1. *Let H be a subgroup of G.*
(i) $\pi_H : K[G] \to K[H]$ *is a $K[H]$-bimodule homomorphism.*
(ii) *If $0 \neq I \triangleleft K[G]$, then $0 \neq \pi_H(I) \triangleleft K[H]$.*
(iii) *If $\alpha \in K[G]$ and X is a right transversal for H in G, then*

$$\alpha = \sum_{x \in X} \pi_H(\alpha x^{-1}) x.$$

We are concerned here with linear identities, that is equations in $K[G]$ of the form $\sum_1^n \alpha_i x \beta_i = 0$ which hold for all $x \in G$ or perhaps for all x in some *large* subset of G. Such identities arise frequently and we will discuss a technique called the Δ-method which has proved successful in handling them. This method was discovered independently by me in 1962 and by M. Smith in 1971. In addition, it is related to methods used in operator algebras and in ergodic theory (see [**2**] for a discussion of the latter). As we will see, there is really no best formulation of this technique. Rather, we will consider a number of variants of it, along with the problems they apply to.

We start with the identity $\alpha x - x\alpha = 0$, $\forall x \in G$, which merely says that $\alpha \in Z(K[G])$, the center of the group algebra. Using $\alpha = x^{-1}\alpha x = \alpha^x$ it follows easily that

LEMMA 2. $\alpha \in Z(K[G])$ if and only if $\alpha = \sum a_x \hat{k}_x$ where k_x is the conjugacy class of $x \in G$ and $a_x \in K$.

But the support of α is finite, so the above conjugacy classes must be of finite size. This motivates us to define
$$\Delta(G) = \{x \in G \mid |k_x| = |G : C_G(x)| < \infty\}$$
and
$$\Delta^+(G) = \{x \in \Delta(G) \mid o(x) < \infty\}$$
where $o(x)$ denotes the order of x. We then have

LEMMA 3. $\Delta(G)$ and $\Delta^+(G)$ are characteristic subgroups of G.
(i) Δ^+ is generated by the finite normal subgroups of G.
(ii) Δ/Δ^+ is torsion-free abelian.
(iii) If $\Delta^+ = 1$, then $K[\Delta]$ is a domain (in fact each nonzero element of $K[\Delta]$ is regular in $K[G]$).

Thus $\Delta = \Delta(G)$ is very well behaved for an infinite group. It is analogous to the center of G and is called the f.c. (finite conjugate) center. Set $\theta = \pi_\Delta$. The key result is then

PROPOSITION 4. Assume that $\sum_i \alpha_i x \beta_i = 0$, $\forall x \in G$. Then for all $x \in G$ we have
(i) $\sum_i \theta(\alpha_i) x \beta_i = 0$,
(ii) $\sum_i \alpha_i x \theta(\beta_i) = 0$,
(iii) $\sum_i \theta(\alpha_i) x \theta(\beta_i) = 0$.

PROOF. We will show at least that $\sum_i \theta(\alpha_i) \beta_i = 0$. If this expression is not zero, fix v in its support and for each i write $\alpha_i = \theta(\alpha_i) + \overline{\alpha}_i$ with $\operatorname{Supp} \overline{\alpha}_i \cap \Delta = \emptyset$. We now proceed in a series of three steps.

Step 1. Set $W = \bigcap C_G(y)$ where the intersection is over all $y \in \operatorname{Supp} \theta(\alpha_i)$ for all i. Then $|G : W| < \infty$ and we will restrict our attention to group elements $x \in W$. Note that any such x centralizes each $\theta(\alpha_i)$.

Step 2. With $x \in W$, multiply the given identity on the left by x^{-1} to obtain $\sum_i \alpha_i{}^x \beta = 0$. This yields $\sum_i \theta(\alpha_i) \beta_i = -\sum_i \overline{\alpha}_i{}^x \beta_i$, and since v is in the support of the left-hand term, it must occur on the right. Thus there exist $a \in \operatorname{Supp} \overline{\alpha}_i$, $b \in \operatorname{Supp} \beta_i$, for some i, with $v = a^x b$. Hence $a^x = vb^{-1}$ and $x \in C_W(a) w_{a,b}$, a fixed right coset of $C_W(a)$ depending upon a, b, and v. This implies that W is the finite union $W = \bigcup_{a,b} C_W(a) w_{a,b}$.

Step 3. We now invoke the following lemma to conclude that $|W : C_W(a)| < \infty$ for some a. But this is a contradiction since $|G : W| < \infty$ and since $a \in \operatorname{Supp} \overline{\alpha}_i$ implies that $|G : C_G(a)| = \infty$. The remainder of the proof is routine.

LEMMA 5. *If G is the finite union of cosets $G = \bigcup_1^n H_i x_i$, then $|G : H_i| \leq n$ for some i.*

As a first application we have

COROLLARY 6. *Let A, B be ideals of $K[G]$ with $AB = 0$. Then $\theta(A)\theta(B) = 0$.*

This reduces the question of whether $K[G]$ is prime or semiprime to the subring $K[\Delta]$ where it is easily solved. Indeed we have

THEOREM 7 [1]. *$K[G]$ is prime if and only if $\Delta^+(G) = 1$ (that is, G has no nontrivial finite normal subgroup).*

If char $K = 0$, it is an easy result that $K[G]$ is semiprime. For char $K = p > 0$ we have

THEOREM 8 [4]. *Let char $K = p > 0$. Then $K[G]$ is semiprime if and only if G has no finite normal subgroup of order divisible by p.*

Now let us look at our first variation on this method. In the proof of Proposition 4, the linear identity was assumed to hold for all $x \in G$, and what we needed was that $G \cap W$ could not be covered by a finite union of cosets of subgroups of infinite index. Other subsets of G also have this property. Let T be a subset of G. We say that T is large if for all subgroups W of finite index in G, $T \cap W$ cannot be covered by a finite union of cosets of subgroups of infinite index. For example, T could be a subgroup of G of finite index. We say that T is very large if T and all its right translates Tx are large. For example, T could be the nonempty complement of a finite union of cosets.

LEMMA 9. *Let T be a very large subset of G. If $\sum_1^n \alpha_i x \beta_i = 0$ for all $x \in T$, then the identity holds for all $x \in G$.*

This can be proved by using the argument of Proposition 4 to deduce that $\sum_i \theta(\alpha_i x^{-1})xy\beta_i = 0$ for all $x, y \in G$. Then we merely sum these expressions over all $x \in X$, a transversal for Δ in G. As a minor application we mention

LEMMA 10. *Let $K[G]$ be prime with $G \neq 1$. If $0 \neq \alpha, \beta \in K[G]$, then for all $n \geq 1$ there exists $\delta_n \in K[G]$ with $|\text{Supp } \alpha \delta_n \beta| \geq n$.*

PROOF. Note that G is infinite so that if $S \subseteq G$ is finite, then $T = G \backslash S$ is a very large subset of G. We proceed by induction on n. Given δ_{n-1} we see that $(\text{Supp } \alpha g \beta) \cap (\text{Supp } \alpha \delta_{n-1} \beta) \neq \emptyset$ implies that $g \in S$, a finite subset of G. If we can find $g \in T = G \backslash S$ with $\alpha g \beta \neq 0$, then clearly $\delta_n = g + \delta_{n-1}$ will suffice. On the other hand, if $\alpha g \beta = 0$ for all $g \in T$, then $\alpha g \beta = 0$ for all $g \in G$ and hence $K[G]$ is not prime, a contradiction.

For a second variant, we restrict the linear identity to hold for all $x \in H$, a subgroup of G. We then define the almost centralizer of H in G to be

$$D_G(H) = \{g \in G | \, |H : C_H(g)| < \infty\}.$$

This is clearly a subgroup of G with $\mathbb{D}_G(H) \cap H = \mathbb{D}_H(H) = \Delta(H)$. The same argument yields, for example,

LEMMA 11. *Let H be a subgroup of G, $D = \mathbb{D}_G(H)$, and let the identity $\sum_i \alpha_i x \beta_i = 0$ hold for all $x \in H$. Then $\sum_i \pi_D(\alpha_i)\beta_i = 0$.*

A very nice application of this is the following intersection theorem.

THEOREM 12 [3]. *Let $H \triangleleft G$ with $\mathbb{D}_G(H) \subseteq H$ and assume that $K[H]$ is prime. If $0 \neq I \triangleleft K[G]$, then $I \cap K[H] \neq 0$.*

PROOF. Choose $0 \neq \alpha \in I$ so that $\mathrm{Supp}\,\alpha$ meets the minimal number, say $n+1$, of cosets of H. We can assume that $\mathrm{tr}\,\alpha \neq 0$ so we can write $\alpha = \sum_{i=0}^n \alpha_i$ with $\alpha_i \in K[Hx_i]$ and $x_0 = 1$. Take $h \in H$ and form

$$\beta = \alpha_0 h \alpha - \alpha h \alpha_0 \in I$$
$$= \sum_0^n (\alpha_0 h \alpha_i - \alpha_i h \alpha_0) = \sum_0^n \beta_i.$$

Since $H \triangleleft G$ and $\alpha_0 \in K[H]$, we have $\beta_i \in K[Hx_i]$. Furthermore, $\beta_0 = 0$ so the minimality of n implies that $\beta_i = 0$ for all i. In particular, if $i \neq 0$ we obtain the linear identity

$$\alpha_0 h \alpha_i - \alpha_i h \alpha_0 = 0 \quad \forall h \in H.$$

Hence $\pi_D(\alpha_0)\alpha_i = \pi_D(\alpha_i)\alpha_0$. But $D \subseteq H$ and $\mathrm{Supp}\,\alpha_i \cap H = \varnothing$ so $\pi_D(\alpha_i) = 0$ and $\pi_D(\alpha_i)\alpha_0 = 0$. Finally, $D = \Delta(H)$ is torsion-free abelian since $K[H]$ is prime, and $\pi_D(\alpha_0) \neq 0$ since $\mathrm{tr}\,\alpha \neq 0$. Thus $\pi_D(\alpha_0)$ is regular in $K[G]$ and we conclude that $\alpha_i = 0$ for all $i \neq 0$.

Recall that $G = \bigcup_1^n H_i x_i$ implies that $|G : H_i| \leq n$ for some i. So far we have only used $|G : H_i| < \infty$. Now we want numerical information which will apply even when G is finite. This leads us to define for each integer $k \geq 1$

$$\Delta_k(G) = \{x \in G \mid |G : \mathbb{C}_G(x)| \leq k\}, \qquad \theta_k \colon K[G] \to K[\Delta_k].$$

Note that Δ_k is a normal subset of G, but not necessarily a subgroup. Furthermore, $\Delta_a \Delta_b \subseteq \Delta_{ab}$.

We would like the linear identity $\sum_i \alpha_i x \beta_i = 0$ to imply $\sum_i \theta_k(\alpha_i)\beta_i = 0$ for a particular computable integer k. To do this, we look again at the proof of Proposition 4. We see that it is necessary to control (1) the index $|G : W|$ and (2) the number of cosets which occur in the union $W = \bigcup_{a,b} \mathbb{C}_W(a) w_{a,b}$. The latter is easy if we know $|\mathrm{Supp}\,\alpha_i|$, $|\mathrm{Supp}\,\beta_i|$ and the number of terms in the sum. In fact, this suffices for both parts as we will see later.

We first discuss an application where (1) is realized by restricting the parameters involved. We study group rings satisfying a polynomial identity. If $K[G]$ satisfies a p.i. of $\deg n$ then by linearization it satisfies

$$f(\varsigma_1, \varsigma_2, \ldots, \varsigma_n) = \sum_{\sigma \in \mathrm{Sym}_n} a_\sigma \varsigma_{\sigma 1} \varsigma_{\sigma 2} \cdots \varsigma_{\sigma n}$$

with $a_\sigma \in K$ and $a_1 = 1$. In particular, if we plug group elements into this expression we get at most $n!$ terms each with support size 1.

Here is a brief idea of the proof (we ignore the subscript k for simplicity). Write
$$f = \varsigma_1 g(\varsigma_2, \ldots, \varsigma_n) + \sum_\lambda \lambda(\varsigma_2, \ldots, \varsigma_n) \varsigma_1 g_\lambda(\varsigma_2, \ldots, \varsigma_n)$$
where $\lambda \neq 1$ is a monomial and g_λ is a polynomial in $\varsigma_2, \ldots, \varsigma_n$. Fix $x_2, \ldots, x_n \in G$ with $\lambda(x_2, \ldots, x_n) \notin \Delta$ for all $\lambda \neq 1$, plug in $\varsigma_i = x_i$ for all $i \geq 2$, and let $\varsigma_1 = x$ in f. This yields
$$0 = xg(x_2, \ldots, x_n) + \sum_\lambda \lambda(x_2, \ldots, x_n) x g_\lambda(x_2, \ldots, x_n)$$
for all $x \in G$ and by the Δ-method we obtain
$$0 = g(x_2, \ldots, x_n) + \sum_\lambda \theta(\lambda(x_2, \ldots, x_n)) g_\lambda(x_2, \ldots, x_n)$$
$$= g(x_2, \ldots, x_n)$$
since $\lambda(x_2, \ldots, x_n) \notin \Delta$. Thus we get a p.i. of smaller degree, but one which is satisfied only for certain $x_2, \ldots, x_n \in G$. By keeping track of all this information and using linear identities on subsets of G we actually obtain

LEMMA 13. *Let $K[G]$ satisfy a p.i. of degree n. Then there exist integers $k = k(n)$ and $t = t(n)$ with $|G : \Delta_k| \leq t$.*

Of course Δ_k is not a subgroup of G in general so $|G : \Delta_k| \leq t$ means that $G = \bigcup_1^t \Delta_k x_i$, a union of t right translates of Δ_k. From this we can deduce the p.i. result. This was known for char $K = 0$ by different means. In characteristic p we have

THEOREM 14 [5, 8]. *Let char $K = p > 0$ and suppose $K[G]$ satisfies a polynomial identity of degree n. Then there exists a subgroup A of G whose commutator subgroup A' is a finite p-group and such that $|G : A| \cdot |A'| \leq g(n)$, where the latter is an appropriate function of n.*

Finally here is a variant and application which is not at all well known. Start with $\sum_i \alpha_i x \beta_i = 0$ for all $x \in G$ and assume that we are given $m = \sum_i (|\operatorname{Supp} \alpha_i| + |\operatorname{Supp} \beta_i|)$. Pick an integer k. Then the Δ-method shows that either $\sum_i \theta_k(\alpha_i) x \beta_i = 0$ or there exists $g \in \operatorname{Supp} \alpha_i$, for some i, with $k < |G : C_G(g)| \leq f(k)$ where the latter function of k depends upon m.

Now take $k_0 = 1$ and inductively define $k_n = f(k_{n-1})$. Then $k_0 < k_1 < \cdots < k_{m+1}$ so there must exist a subscript $t \leq m$ with
$$\left(\bigcup_i \operatorname{Supp} \alpha_i \right) \cap (\Delta_{k_{t+1}} \setminus \Delta_{k_t}) = \emptyset.$$
But then we have $\sum \theta_{k_t}(\alpha_i) x \beta_i = 0$. We have therefore proved

LEMMA 15. *Given the linear identity $\sum_i \alpha_i x \beta_i = 0$ for $x \in G$, set $m = \sum_i(|\operatorname{Supp} \alpha_i| + |\operatorname{Supp} \beta_i|)$. Then there exists $k \leq h(m)$ with $\sum_i \theta_k(\alpha_i) x \beta_i = 0$, $\forall x \in G$. Here h is a fixed function of m.*

This can be easily extended to deal with products of more terms. For example we have

LEMMA 16. *Given $\alpha_0, \alpha_1, \ldots, \alpha_n \in K[G]$ with*

$$\alpha_0 x_1 \alpha_1 x_2 \cdots x_n \alpha_n = 0 \qquad \forall x_1, x_2, \ldots, x_n \in G.$$

Then there exists an integer k bounded by a function of $\sum_i |\operatorname{Supp} \alpha_i|$ with

$$\theta_k(\alpha_0) x_1 \theta_k(\alpha_1) x_2 \cdots x_n \theta_k(\alpha_n) = 0 \qquad \forall x_1, x_2, \ldots, x_n \in G.$$

As an application we mention the following technical result used to study the Jacobson radical of group rings of locally solvable groups.

PROPOSITION 17 [7]. *Let $G = QH$ with $Q \triangleleft G$ and $|H| \leq t < \infty$. Assume that $JK[Q] = 0$ and that $\alpha \in JK[G] \cap K[H]$. Then there exists a subgroup $W \triangleleft H$ with*
  (i) $|Q : C_Q(w)| \leq g(t)$, $\forall w \in W$,
  (ii) $\pi_W(\alpha) \in JK[G] \cap K[W]$.
Here g is a fixed function of t.

PROOF. Observe that $|G : Q| \leq |H| \leq t$. Since $JK[Q] = 0$, it now follows (see [6, Theorem 7.2.7]) that $JK[G]^t = 0$. Thus, since $\alpha \in JK[G]$ we have

$$\alpha x_1 \alpha x_2 \cdots \alpha x_{t-1} \alpha = 0 \qquad \forall x_1, x_2, \ldots, x_{t-1} \in G.$$

Furthermore, $\alpha \in K[H]$ implies that $|\operatorname{Supp} \alpha| \leq |H| \leq t$. Thus Lemma 16 applies and there exists an appropriate integer k with

$$\theta_k(\alpha) x_1 \theta_k(\alpha) x_2 \cdots x_{t-1} \theta_k(\alpha) = 0 \qquad \forall x_1, x_2, \ldots, x_{t-1} \in G,$$

and therefore $\theta_k(\alpha) \in JK[G]$. Now let W be the normal closure in H of the support of $\theta_k(\alpha)$. A more careful choice of k will guarantee that $\theta_k(\alpha) = \pi_W(\alpha)$.

The material discussed here was all done before 1979. But the Δ-method (as we will see) is still being used in crossed products and in the computation of certain rings of quotients. It has been an important technique for almost 25 years and it will certainly continue to be useful in the future.

References

1. I. G. Connell, *On the group ring*, Canad. J. Math. **15** (1963), 650–685.
2. D. R. Farkas, *Recurrent behavior in rings* (to appear).
3. E. Formanek and A. I. Lichtman, *Ideals in group rings of free products*, Israel J. Math. **31** (1978), 101–104.
4. D. S. Passman, *Nil ideals in group rings*, Michigan Math. J. **9** (1962), 375–384.
5. ____, *Group rings satisfying a polynomial identity*, J. Algebra **20** (1972), 103–117.
6. ____, *The Algebraic Structure of Group Rings*, Wiley-Interscience, New York, 1977.
7. ____, *The Jacobson radical of the group ring of a locally solvable group*, Proc. London Math. Soc. (3) **38** (1979), 169–192; Corrigendum, **39** (1979), 208–210.
8. M. K. Smith, *Group algebras*, J. Algebra **18** (1971), 477–499.

2. The Jacobson Radical of Group Rings

The last section concerned methods; this one will be speculation on a specific question. We wish to determine when $K[G]$ is semiprimitive (equivalently semisimple) and more generally to describe the Jacobson radical $JK[G]$. This is my favorite problem. It was suggested to me by my advisor, R. Brauer, who learned about its infinite aspects from I. Kaplansky (see [4]) at the famous Ram's Head Inn conference. For finite groups the result is easy.

THEOREM 1 [5]. *Let G be finite. Then $K[G]$ is semiprimitive if and only if $|G| \neq 0$ in K.*

From this it appears that for all groups G, if char $K = 0$ then $JK[G] = 0$. The best result here is proved by studying the behavior of the radical under field extensions.

THEOREM 2 [1]. *Let A be an algebra over the field F and let $K \supseteq F$ be a field extension.*
(i) *If K/F is separably algebraic, then $J(K \otimes A) = K \otimes (JA)$.*
(ii) *If K/F is purely transcendental and $K \neq F$, then $J(K \otimes A) = K \otimes I$ where I is a nil ideal of A.*

Since $K \otimes F[G] = K[G]$, the above applies to yield

COROLLARY 3 [2]. *Let char $K = 0$ with K not algebraic over the rationals. If G is any group, then $K[G]$ is semiprimitive.*

PROOF. Since K/Q is not algebraic we have $K \supseteq F \supseteq Q$ where F/Q is purely transcendental, $F \neq Q$ and K/F is separably algebraic. Thus it suffices to show that $Q[G]$ has no nonzero nil ideals. For this, let $*\colon Q[G] \to Q[G]$ be the natural involution determined by $x^* = x^{-1}$ for all $x \in G$. If $\alpha = \sum a_x x \neq 0$, then $\alpha\alpha^*$ is $*$-symmetric with trace $\sum a_x^2 \neq 0$. Therefore if $0 \neq \alpha \in I \triangleleft Q[G]$, then $0 \neq \alpha\alpha^* \in I$ and $(\alpha\alpha^*)^{2^n} \neq 0$ for all n so I is not nil.

Thus in characteristic 0, this leaves only the rationals Q and (equivalently) its algebraic extensions. For these fields, a number of special classes of groups have been considered and in each such case $JQ[G] = 0$. But there has been no real progress for 25 years. Logicians have tried model theoretic techniques but also to no avail. The semiprimitivity of $Q[G]$ is still a wide open problem.

In characteristic $p > 0$ one can follow the above lead for p'-groups, that is groups with no elements of order p. This assumption implies that $K[G]$ has no nonzero nil ideals. Also one knows that any finitely generated field extension of $GF(p)$ is necessarily separably generated. Thus we have

COROLLARY 4. *Let* char $K = p > 0$ *with K not algebraic over $GF(p)$. If G is a p'-group, then $K[G]$ is semiprimitive.*

It is not true that $K[G]$ semiprimitive implies G is a p'-group. The first example, due to Wallace, is G infinite dihedral in characteristic 2. Another example is the wreath product $G = Z \operatorname{wr} Z_p$ in characteristic p where Z denotes the infinite cyclic group and $|Z_p| = p$. In fact if $G = Z_p \operatorname{wr} Z$ or $\operatorname{Sym}_\infty$ (the countably infinite, locally finite symmetric group), then $K[G]$ is a primitive ring in characteristic p.

Let us study the characteristic p situation in more detail. Let R be any ring and observe that JR contains all nil and nilpotent ideals of R. Thus if NR, the nilpotent radical of R, is defined to be the sum of all nilpotent ideals of R, then NR is a characteristic nil ideal of R contained in JR. For a group G we set

$$\Delta^p(G) = \langle x \in \Delta^+(G) | x \text{ is a } p\text{-element}\rangle,$$

that is Δ^p is generated by all elements of Δ^+ of order a power of p. We have the following Δ-method result.

THEOREM 5 [6]. *Let* char $K = p > 0$.
(i) $NK[G] = JK[\Delta^p] \cdot K[G]$.
(ii) $JK[\Delta^p] = \bigcup_W JK[W]$ *where W runs over all finite normal subgroups of G generated by p-elements.*
(iii) $NK[G]$ *is nilpotent if and only if Δ^p is finite.*

Part (i) above is an example of controlling. Let $I \triangleleft K[G]$ and let $H \triangleleft G$. We say that H controls I if $I = (I \cap K[H]) \cdot K[G]$, that is $K[H]$ contains generators for I as an ideal. It is easy to show that for each ideal I there exists a unique smallest normal subgroup $C = \mathcal{C}(I)$, the controller of I, so that $H \triangleleft G$ controls I if and only if $H \supseteq C$.

With the assumptions of Theorem 5 we can also show that $\mathcal{C}(NK[G]) = \Delta^p(G)$. By accident, the above result might prove to be important here. This is because in all cases so far computed it has been observed that G finitely generated implies $JK[G] = NK[G]$. We remark that there exist many computations here due to Villamayor, Wallace, Zalesskii (and his student, Dyment) and me (and my student, Hampton). We will not discuss these methods here, and in so doing we are leaving out the bulk of published material on the Jacobson radical of $K[G]$. Instead we will speculate as to the correct answer based on these computations.

LEMMA 6. *Let $H \subseteq G$.*
(i) *If W is an irreducible $K[H]$-module, then there exists an irreducible $K[G]$-module V such that W is a submodule of the restriction V_H.*

(ii) $JK[G] \cap K[H] \subseteq JK[H]$.

PROOF. (i) Take $W = K[H]/M$ where M is a maximal right ideal of $K[H]$. By freeness we have $M \cdot K[G] \neq K[G]$ so there exists a maximal right ideal L of $K[G]$ with $L \supseteq M \cdot K[G]$. It follows that $L \cap K[H] = M$ so

$$W = K[H]/M = K[H]/(L \cap K[H]) \subseteq K[G]/L = V.$$

(ii) Let $\alpha \in JK[G] \cap K[H]$ and let W be an irreducible $K[H]$-module. Then $W \subseteq V_H$ and $V\alpha = 0$ so $W\alpha = 0$ and $\alpha \in JK[H]$.

It follows from (ii) above that the property of belonging to $JK[G]$ is *local*, that is it depends on finitely generated subgroups of G. Because of this we need a local version of Δ and Δ^+. We define

$$\Lambda(G) = \{x \in G | x \in \Delta(H) \text{ for all finitely generated}$$
$$\text{subgroups } H \text{ containing } x\},$$
$$\Lambda^+(G) = \{x \in \Lambda(G) | o(x) < \infty\}.$$

These also have fairly nice properties, one of which is quite surprising.

LEMMA 7. *$\Lambda(G)$ and $\Lambda^+(G)$ are characteristic subgroups of G.*
(i) *Λ^+ is locally finite.*
(ii) *Λ/Λ^+ is torsion-free abelian.*
(iii) *If $H \triangleleft G$ with $H \subseteq \Lambda^+$, then $\Lambda(G/H) = \Lambda(G)/H$ and $\Lambda^+(G/H) = \Lambda^+(G)/H$.*

Part (iii) above says that Λ^+ is a *radical*. This is not true of $\mathbb{Z}, \Delta^+, \Delta$ or Λ. Also if G is a locally finite group, then $G = \Lambda^+(G)$ so Λ^+ can be an arbitrary locally finite group.

We now need a local version of the nilpotent radical NR. Let R be a ring and set

$$N^*R = \{r \in R | rS \text{ is nilpotent for all finitely}$$
$$\text{generated subrings } S \subseteq R\}.$$

It is easy to see that N^*R is a characteristic nil ideal of R. Furthermore if R is a K-algebra then

$$N^*R = \{r \in R | rS \text{ is nilpotent for all finitely}$$
$$\text{generated subalgebras } S \subseteq R\}.$$

Note that if R is a finitely generated K-algebra then $N^*R = NR$. The following result is, for the most part, an immediate corollary of Theorem 5.

THEOREM 8 [7]. *Let $K[G]$ be given with char $K = p > 0$.*
(i) *If $JK[H] = NK[H]$ for all finitely generated subgroups $H \subseteq G$, then $JK[G] = N^*K[G]$.*
(ii) *$N^*K[G] = JK[\Lambda^+] \cdot K[G]$.*
(iii) *$N^*(K[G]/N^*K[G]) = 0$.*

Thus, by (ii), $\Lambda^+(G)$ controls $N^*K[G]$. While it is not the controller in general, we do have $\mathcal{C}(N^*K[G]) \subseteq \Lambda^+(G)$ so that the controller is a locally finite characteristic subgroup of G. Part (iii) above says that N^* is a *radical* for group rings even though it is not a radical for rings in general. This lends additional support to the conjecture that $JK[G] = N^*K[G]$ for all groups G. In view of (i) above this is equivalent to showing that $JK[G] = NK[G]$ for all finitely generated groups G. Frankly it is hard to use this extra hypothesis (that G is finitely generated). In the computations so far, this assumption was sometimes used to deal with certain aspects of the Burnside problem which occur because of limitations in the available techniques. One test problem I propose is: Determine $JK[G]$ for G finitely generated and residually finite.

Let us move on. Because of (ii), or for its own intrinsic interest, we certainly want to settle the case of locally finite groups. Here a version of subnormality comes into play. Let G be locally finite. A finite subgroup A is said to be locally subnormal in G (written $A \operatorname{lsn} G$) if $A \triangleleft \triangleleft B$ for all finite B with $G \supseteq B \supseteq A$. It is a result of [3] that for A finite, $A \operatorname{lsn} G$ if and only if A is serial (that is there exists a generalized series from A to G).

LEMMA 9. *Let G be a locally finite group and let* $\operatorname{char} K = p > 0$.
(i) *If* $A \operatorname{lsn} G$, *then* $JK[A] \subseteq JK[G]$.
(ii) *If G is solvable, then $JK[G] \neq 0$ if and only if there exists $A \operatorname{lsn} G$ with* $p \mid |A|$.

PROOF. (i) It suffices to show that $JK[A] \cdot K[G]$ is a nil right ideal. Since this is a local property, we need only check that $JK[A] \cdot K[B]$ is nil for all finite $B \supseteq A$. But $A \triangleleft \triangleleft B$ and $|B| < \infty$ so it follows easily that $JK[A] \subseteq JK[B]$ thereby yielding the result.

(ii) If A exists then $JK[G] \neq 0$ by (i). Conversely assume $JK[G] \neq 0$ and let $1 = G_0 < G_1 < \cdots < G_n = G$ be the derived series for the solvable group G. Since $JK[G_0] = 0$ and $JK[G_n] \neq 0$ we can choose $k \geq 0$ maximal with $JK[G_k] = 0$ and have $k < n$. Thus $JK[G_{k+1}] \neq 0$ so there exists a group H with $G_k \subseteq H \subseteq G_{k+1}$, H/G_k finite and $JK[G_{k+1}] \cap K[H] \neq 0$. Now $JK[H] \neq 0$ and $|H/G_k| < \infty$ imply that $JK[H]$ is nilpotent. Thus there exists a finite normal subgroup A of H of order divisible by p. Since $A \triangleleft H, H \triangleleft G_{k+1}$ and $G_{k+1} \triangleleft G$ we have $A \triangleleft \triangleleft G$.

Let G be locally finite. In view of the preceding result we define

$$\int(G) = \langle A \mid A \operatorname{lsn} G \rangle.$$

Thus $S = \int(G)$ is a group which is generated by its locally subnormal subgroups. Possible examples of such groups are (i) S a p-group or (ii) S an f.c. group, that is $S = \Delta(S)$. As we will see these examples are the building blocks for a general S. For each integer $n \geq 1$ define

$$\int_n(G) = \langle A \mid A \operatorname{lsn} G \text{ and } l(A) \leq n \rangle$$

where $l(A)$ denotes the composition length of A. For any operator W on groups we set
$$W^p(G) = \langle x \in W(G) | x \text{ is a } p\text{-element} \rangle.$$
This is of course consistent with our definition of Δ^p.

THEOREM 10 [8]. *Let G be a locally finite group and p a prime. If $O_p(G) = 1$, then $\int^p(G)$ is the ascending set theoretic union $\int^p(G) = \bigcup_{n=1}^{\infty} \int_n^p(G)$ of the normal f.c. groups $\int_n^p(G)$.*

Note that $\int^p(G)$ need not be an f.c. group in the above situation. Now let us add the normal p-subgroup $O_p(G)$ to the structure. We define $\mathbb{S}_p(G) \supseteq O_p(G)$ by
$$\mathbb{S}_p(G)/O_p(G) = \int^p(G/O_p(G)).$$
Thus $\mathbb{S}_p$ is the extension of a normal p-subgroup by an ascending union of normal f.c. groups. Based on the few computations with locally finite groups, it appears that $\mathbb{S}_p(G)$ might be the controller of $JK[G]$. For this reason we consider the ideal I below.

THEOREM 11 [9]. *Let G be a locally finite group and let $\operatorname{char} K = p > 0$. Set $I = JK[\mathbb{S}_p(G)] \cdot K[G]$.*
  (i) *I is a nil ideal of $K[G]$ so $I \subseteq JK[G]$.*
  (ii) *I is a semiprime ideal, that is $K[G]/I$ is a semiprime ring.*
  (iii) *If $\Delta(G) = 1$, then I is a prime ideal.*

Thus we have at least some support for the conjecture that if G is locally finite and $\operatorname{char} K = p > 0$, then
$$JK[G] = JK[\mathbb{S}_p(G)] \cdot K[G].$$
Note that if this formula holds then we necessarily have $\mathbb{S}_p(G) = \mathcal{C}(JK[G])$.

But the latter controller statement seems somewhat dubious since $\mathbb{S}_p$ is not a radical operator, that is we do not have $\mathbb{S}_p(G/\mathbb{S}_p(G)) = 1$ in general. Thus we look at the $\int$-radical of G. If W is a group, we say that W is hyper-$\int$ if every nonidentity homomorphic image $\overline{W}$ of W satisfies $\int(\overline{W}) \neq 1$.

LEMMA 12. *Let G be a locally finite group and let $H = \mathsf{H}(G)$ be generated by all the hyper-$\int$ normal subgroups of G.*
  (i) *H is the unique largest hyper-$\int$ normal subgroup of G.*
  (ii) *$\mathsf{H}(G/H) = \int(G/H) = 1$.*
  (iii) *$H \supseteq \mathbb{S}_p(G)$ for any prime p.*

Thus H is a radical and we discover

COROLLARY 13 [9]. *Let G be locally finite and let $\operatorname{char} K = p > 0$. If $H = \mathsf{H}(G)$ and $S = \mathbb{S}_p(G)$, then $JK[H] = JK[S] \cdot K[H]$.*

This at least eliminates an obvious objection to the conjecture. But the question is still wide open. Two problems of particular interest here are: (1) Study linear groups in prime characteristic $q \neq p$. Perhaps just look at the special case of Chevalley groups over infinite algebraic extensions of $GF(q)$. (2) Assume that the orders of the finite p-subgroups of G are bounded. Does this imply that $JK[G]$ is nilpotent? The conclusion would follow if the Cartan invariants of a finite group W were bounded by a function of the order of the Sylow p-subgroups of W.

It now appears that the study of $JK[G]$ with G locally finite will require a very close look at the Jacobson radicals of group rings of finite groups. However it is not clear just what information we will need from the latter. We mention two results on finite groups which are somewhat suggestive. The first offers a description of $JK[G]$ for G a p-solvable group. The second shows that (2) holds at least for the radical of the center of $K[G]$.

THEOREM 14 [11]. *Let* $\operatorname{char} K = p > 0$, *let* G *be finite and define the p-trace* $t_p \colon K[G] \to K$ *by*
$$t_p\left(\sum_{x \in G} a_x x\right) = \sideset{}{'}\sum a_x$$
where the latter sum $\sum'$ is over all p-elements of G.
(i) $t_p(JK[G]) = 0$.
(ii) *If G is p-solvable, then $JK[G]$ is the largest ideal I of $K[G]$ with $t_p(I) = 0$.*

We note that the p-elements of G are precisely the elements whose p'-part is trivial. More generally we can consider those elements of G whose p'-part is contained in a fixed conjugacy class of G. Then each of these p-sections of G gives rise to a trace map $K[G] \to K$ and by considering all of these traces, we obtain a description of $JK[G]$ as in (ii) which applies to all finite groups.

THEOREM 15 [10]. *Let* $\operatorname{char} K = p > 0$ *and let G be a finite group with Sylow p-subgroup P. Then $J(\mathbb{Z}(K[G]))$ is nilpotent of degree bounded by a function of $|P|$.*

Finally we return to arbitrary groups G. By combining the two conjectures we have proposed, we obtain as a possible description of the Jacobson radical the expression
$$JK[G] = JK[\mathbb{S}_p(\Lambda^+(G))] \cdot K[G].$$
Furthermore, if this holds then $\mathbb{S}_p(\Lambda^+(G)) = \mathcal{C}(JK[G])$ and $JK[G] = 0$ if and only if $\mathbb{S}_p(\Lambda^+(G)) = 1$. Thus the description would be sufficiently sharp to solve the semiprimitivity problem in characteristic p.

We close by tabulating some classes of groups for which the conjecture is known to hold. These are the so-called computations we have alluded to throughout this section. The results appear in a sequence of papers by Zalesskii, Hampton and myself.

THEOREM 16. *Let* $\operatorname{char} K = p > 0$. *Then*
$$JK[G] = JK[\mathbb{S}_p(\Lambda^+(G))] \cdot K[G]$$

for G a locally solvable group or a linear group over a field F of characteristic 0 or p.

In addition, if G is a linear group over a field F of characteristic $q \neq p$, then we know at least that $JK[G] = N^*K[G]$ and that the conjecture holds for $G = \mathrm{SL}_n(F)$ or $\mathrm{GL}_n(F)$. Unfortunately, there has been no real progress on this conjecture in 7 years.

References

1. S. A. Amitsur, *The radical of field extensions*, Bull. Res. Council Israel **7F** (1957), 1–10.
2. ____, *On the semi-simplicity of group algebras*, Michigan Math. J. **6** (1959), 251–253.
3. B. Hartley, *Serial subgroups of locally finite groups*, Proc. Cambridge Philos. Soc. **71** (1972), 199–201.
4. I. Kaplansky, *"Problems in the theory of rings" revisited*, Amer. Math. Monthly **77** (1970), 445–454.
5. H. Maschke, *Über den arithmetischen Charakter der Coefficienten der Substitutionen endlicher linearer Substitutionsgruppen*, Math. Ann. **50** (1898), 482–498.
6. D. S. Passman, *Radicals of twisted group rings*, Proc. London Math. Soc. (3) **20** (1970), 409–437.
7. ____, *A new radical for group rings?*, J. Algebra **28** (1974), 556–572.
8. ____, *Subnormality in locally finite groups*, Proc. London Math. Soc. (3) **28** (1974), 631–653.
9. ____, *Radical ideals in group rings of locally finite groups*, J. Algebra **33** (1975), 472–497.
10. ____, *The radical of the center of a group algebra*, Proc. Amer. Math. Soc. **78** (1980), 323–326.
11. Y. Tsushima, *Some notes on the radical of a finite group ring*, Osaka J. Math. **15** (1978), 647–653.

3. Zero Divisors in Group Rings

We continue studying the group algebra $K[G]$ and consider a problem of a different sort. If $1 \neq x \in G$ with $n = o(x) < \infty$, then
$$(1-x)(1+x+\cdots+x^{n-1}) = 1 - x^n = 0$$
and $K[G]$ has zero divisors. On the other hand, if G is torsion-free, then $K[G]$ has no obvious zero divisors. Based on this observation, and with frankly very little supporting evidence, it was conjectured that G torsion-free implies that $K[G]$ is a domain (that is, has no zero divisors). Even so, the conjecture has held up for over 30 years and it is beginning to appear now that it is in fact correct.

In the early days, the work on this problem was for the most part group theoretic in nature and quite unsatisfactory. The idea was to show that G torsion-free implies that G has certain other properties which might trivially imply that $K[G]$ is a domain. Among the special families considered were ordered, right ordered and unique product groups. But certainly most torsion-free groups are not right ordered and it is quite possible that they are not unique product either.

When it finally became time to get one's hands dirty, it was clear that the finitely generated torsion-free abelian-by-finite groups were the groups to consider. Let G be such a group with A a normal abelian subgroup of finite index. If G/A is cyclic, then G is necessarily right ordered so nothing new occurs. The first case of interest is then $G/A = Z_2 \times Z_2$ and here one discovers the nasty group
$$H = \langle x, y | x^{-1}y^2 x = y^{-2},\ y^{-1}x^2 y = x^{-2} \rangle.$$
Then H is torsion-free, $A = \langle x^2, y^2, (xy)^2 \rangle$ is free abelian of rank 3 and H/A is a fours group. Furthermore, H is not right ordered. Ad hoc attacks on the zero divisor problem for this seemingly easy group failed miserably. The question was finally settled because H is supersolvable.

We recall that a group G is supersolvable if it has a normal series $1 = G_0 \subseteq G_1 \subseteq \cdots \subseteq G_k = G$ with each quotient G_{i+1}/G_i cyclic. Some examples are $G = Z$, $Z_2 * Z_2$, the infinite dihedral group and, of course, the group H above.

The first real result on the zero divisor problem was based on work of P. M. Cohn. Let n be a positive integer. A ring R is said to be an n-fir if and only if all n-generator right ideals of R are free of unique rank. This concept is of interest since a 1-fir is precisely a domain.

THEOREM 1 [3]. *Let R_1 and R_2 be n-firs containing a common division ring D. Then the coproduct $R_1 \amalg R_2$ over D is also an n-fir.*

Lewin observed that the division ring D above could be replaced by an Ore domain via localization and then Formanek used it to prove

THEOREM 2 [8]. *If G is torsion-free supersolvable, then $K[G]$ is a domain.*

It is not true that all finitely generated abelian-by-finite groups are supersolvable, but they are contained in the following family. A group G is said to be polycyclic-by-finite if it has a subnormal series $1 = G_0 \triangleleft \cdots \triangleleft G_k = G$ with each quotient G_{i+1}/G_i either infinite cyclic or finite. Note that such groups G are Noetherian and that $K[G]$ is a right and left Noetherian ring. Furthermore, if G is torsion-free, then $K[G]$ is certainly a prime ring so the Goldie machinery is available for its study. Formanek obtained another result which proved to be quite important.

THEOREM 3 [7]. *Let G be a torsion-free Noetherian group and let $\operatorname{char} K = 0$. Then $K[G]$ has no nontrivial idempotents.*

PROOF (SKETCH). We first study $F[G]$ with $\operatorname{char} F = p > 0$. For any $x \in G$ define $\operatorname{tr}_x \colon F[G] \to F$ by

$$\operatorname{tr}_x \left(\sum a_g g \right) = \sum_{g \sim x} a_g$$

where $g \sim x$ indicates that the elements are conjugate in G. Then $\operatorname{tr}_x(\alpha\beta) = \operatorname{tr}_x(\beta\alpha)$ and

$$\operatorname{tr}_x(\alpha^{p^n}) = \sum_{y^{p^n} \sim x} \operatorname{tr}_y(\alpha)^{p^n}.$$

Now let $\bar{e} \in F[G]$ be an idempotent. Then $\bar{e}^{p^n} = \bar{e}$ implies that

$$\operatorname{tr}_x(\bar{e}) = \sum_{y^{p^n} \sim x} \operatorname{tr}_y(\bar{e})^{p^n}$$

for all n. Note that $\operatorname{Supp} \bar{e}$ is finite so $\operatorname{tr}_y(\bar{e}) \neq 0$ for only finitely many conjugacy classes. Thus $\operatorname{tr}_x(\bar{e}) \neq 0$ implies $x^{p^n} \sim x$ for some $n > 0$ and then, since G is Noetherian, we have $x = 1$.

Now if $\operatorname{aug} \colon F[G] \to F$ is the augmentation map, then $\operatorname{aug}(\bar{e})$ is an idempotent in F and hence equal to 0 or 1. We conclude from this, and from the fact that $\operatorname{tr}_x(\bar{e}) = 0$ for all $x \neq 1$, that $\operatorname{tr}_1(\bar{e}) = \operatorname{aug}(\bar{e}) = 0$ or 1. Finally we study $e \in K[G]$ with $\operatorname{char} K = 0$. Using specializations from K to various fields of prime characteristic, we deduce that $(\operatorname{tr} e)^2 = \operatorname{tr} e$ so $\operatorname{tr} e = 0$ or 1. A characteristic 0 result of Kaplansky now implies that $e = 0$ or 1.

We note that in the above characteristic p argument, one only needs to assume that G has no p'-torsion. This trace argument is one of the two ways we know of using the torsion-free hypothesis. The other one is related but more mysterious. If $e \in K[G]$ is an idempotent, then $eK[G]$ is a projective $K[G]$-module. The above result then implies that if $K[G] = P \oplus Q$, the sum of two $K[G]$-modules, then $P = 0$ or $K[G]$.

Projective modules show up in certain homological considerations. Let R be a ring and let M be an R-module. Then M has finite homological dimension if there exists a long exact sequence
$$0 \to P_n \to P_{n-1} \to \cdots \to P_0 \to M \to 0$$
with each P_i projective. We say that R has finite global dimension if all R-modules have finite homological dimension. In case $R = K[G]$ is a group algebra, we need only test the principal module $M = K$ (where all $x \in G$ act like 1).

THEOREM 4 [12]. *Let $|G : H| < \infty$ and assume that $K[H]$ has finite global dimension. If G is torsion-free, then $K[G]$ also has finite global dimension.*

PROOF (SKETCH). We are given the long exact sequence
$$0 \to P_n \to P_{n-1} \to \cdots \to P_0 \to K \to 0$$
of $K[H]$-modules with each P_i projective. We want to lift this to $K[G]$-modules. Ordinary induction of modules, however, will not work since K induces to K^G a module of degree $t = |G : H|$, and this is not the module K that we need. The trick is to use tensor induction since $K \otimes K \otimes \cdots \otimes K = K$. One shows that there exists a $K[G]$-module structure on
$$Q_m = \sum_{i_1 + i_2 + \cdots + i_t = m} P_{i_1} \otimes P_{i_2} \otimes \cdots \otimes P_{i_t}$$
and $K[G]$-module homomorphisms so that
$$\cdots \to Q_k \to Q_{k-1} \to \cdots \to Q_0 \to K \to 0$$
is exact. Roughly speaking, G acts on Q_m by permuting the tensor factors via signed permutations on the cosets of H. Since Q_m is eventually 0, the goal is to show that each Q_m is projective.

If G permutes the set Ω, then $K[\Omega]$, the K-vector space with basis Ω, becomes a special $K[G]$-module which we call a permutation module. Now suppose, for example, that each P_i is free and hence a permutation module for $K[H]$ with H acting semiregularly on the basis (that is, the stabilizer H_α in H of each point α is trivial). It follows that each Q_m is a (signed) permutation module for $K[G]$ with H again acting semiregularly. But then $G_\alpha \cap H = H_\alpha = 1$ so each G_α is finite. Hence, since G is torsion-free we have $G_\alpha = 1$, and this implies that each Q_m is a free $K[G]$-module.

It turns out that the above proof only requires G to have no p-torsion if $\operatorname{char} K = p$, and then we have

COROLLARY 5. *Let G be a polycyclic-by-finite group with no p-torsion and let $\operatorname{char} K = p > 0$. Then $K[G]$ has finite global dimension.*

What good is any of this? Here is a result which should give you an idea. Let P be a projective R-module. Then P is said to be stably free if $P \oplus F_1 = F_2$

with F_1 and F_2 finitely generated free R-modules. This of course implies that P must be finitely generated.

THEOREM 6 [13]. *Suppose R is a ring wich satisfies*
(i) *R is right Noetherian and semiprime,*
(ii) *R has finite global dimension,*
(iii) *all finitely generated projective R-modules are stably free.*
Then R is a domain.

PROOF (SKETCH). Since R is Noetherian we can work with finitely generated R-modules. Note that finite global dimension implies that all R-modules are *close to* the projectives and (iii) says that all projectives are *close to* the free modules and hence to R_R. In particular we can prove that if M is a nonsingular R-module, then the uniform dimension of R divides u.d. M. Furthermore, (i) implies that R_R is nonsingular and hence so are all right ideals. If $I \neq 0$ is such a right ideal, then $0 < $ u.d. $I \leq $ u.d. R and the divisibility yield u.d. $I = $ u.d. R so I is essential in R. Thus u.d. $R = 1$ and, since R is semiprime, the classical ring of quotients of R is a division ring.

It was Brown who first saw how to apply these ideas. He used a localization theorem of Roseblade to embed $K[G]$ in a local ring (so all finitely generated projective modules are free). Then Serre's result, Corollary 5, and the above showed that if G is a torsion-free, abelian-by-(finite p) group in characteristic p, then $K[G]$ is a domain. This was then lifted to characteristic 0 to settle the case of torsion-free abelian-by-finite groups.

Farkas and Snider carried this further with the following two key ideas. (1) The difficulty in applying Theorem 6 directly to $R = K[G]$ is that we do not know whether G torsion-free suffices to imply hypothesis (iii). But G does have a poly-(infinite cyclic) subgroup H of finite index, and all finitely generated projective modules for $S = K[H]$ are stably free. Thus the first idea is to study the finitely generated nonsingular R-modules M and to somehow show that their restriction to S satisfies u.d. $R_S |$ u.d. M_S. In particular, if I is a nonzero right ideal of R, then this divisibility along with $0 < $ u.d. $I_S \leq $ u.d. R_S will yield I_S ess R_S and hence certainly I_R ess R_R. (2) The second idea is to study the finitely generated projective $K[G]$-modules via trace maps. Given such a module P, there exists a complementary module Q with $P \oplus Q = F = K[G]^n$, the free $K[G]$-module of rank n. Then the projection map $e \colon F \to P \subseteq F$ is an idempotent in $\text{End}_{K[G]} F = M_n(K[G])$. Now let $\text{Tr} \colon M_n(K[G]) \to K$ be the composition of the matrix trace followed by the group ring trace. Then $\text{Tr } e$ turns out to be an invariant of P and one computes as in the proof of Theorem 3.

This yields the characteristic 0 part of the following result. Cliff then lifted the trace computations from characteristic p fields to the p-adics to obtain the characteristic p part.

THEOREM 7 [**1, 2, 5**]. *If G is a torsion-free polycyclic-by-finite group, then $K[G]$ is a domain.*

There now exist slightly easier proofs in characteristic 0. But this combination of ideas is still required for the general case. It is interesting to see how the torsion-free assumption is used in the above proof and for this it is best to consider characteristic $p > 0$. Then the Formanek trace argument uses no p'-torsion and the Serre argument uses no p-torsion. Thus the two techniques are both required and nicely complement each other. The following result settles the idempotent problem in characteristic p.

THEOREM 8 [2]. *Let* $\operatorname{char} K = p > 0$ *and let* G *be a polycyclic-by-finite group with no* p'-*torsion. Then* $K[G]$ *has no nontrivial idempotents.*

Theorem 7 is certainly the best result to date on the zero divisor problem. The next case to consider is probably the torsion-free solvable groups. Here $K[G]$ is no longer Noetherian, but sometimes it can be suitably localized and made Noetherian. This approach was used by Snider to obtain some interesting special cases. Still, we have a long way to go and the zero divisor problem is wide open. In addition, there are other questions for arbitrary groups which we have not even mentioned. For example:

a. If G is torsion-free, are all units of $K[G]$ trivial, that is, of the form kg with $k \in K \backslash 0$, $g \in G$? This is not even known for supersolvable groups.

b. If $K[G]$ is a domain, is it necessarily embeddable in a divison ring? This is clear in the Noetherian case and it is known to be true if G is ordered or a 1-relator group.

Now back to polycyclic-by-finite groups. The above proof leaves a number of problems in its wake. Specifically these are:

1. If G is a torsion-free polycyclic-by-finite group, are all the finitely generated projective modules stably free?

2. Let G be a torsion-free polycyclic-by-finite group. Discuss the nature of the division ring of fractions $Q(K[G])$.

3. Let G be a polycyclic-by-finite group with $\Delta^+(G) = 1$. Determine the Goldie rank of $K[G]$.

For the remainder of this section, we briefly indicate what is known about each of the above.

(1) This seems to be the province of Farrell and Hsiang and the methods are topological. Their latest result is

THEOREM 9 [6]. *If* G *is a torsion-free polycyclic-by-finite group, then all finitely generated projective modules for the integral group ring* $Z[G]$ *are stably free.*

Note that once G becomes nonabelian, these stably free modules are not in general free.

(2) Let us denote this division ring of fractions by $K(G)$. We will mention just a few results of interest.

THEOREM 10 [4]. *Let G and H be torsion-free polycyclic-by-finite groups. If G and H are nilpotent, then $K(G) \simeq K(H)$ implies that $G \simeq H$. If G and H are not nilpotent, then this is not necessarily true.*

A counterexample for the latter is constructed as follows. Let $G = A \times_\sigma Z$ and $H = A \times_\tau Z$ where $A = \langle a, b \rangle$ is free abelian of rank 2 and where $\sigma = \begin{pmatrix} 1 & 0 \\ 0 & -1 \end{pmatrix}$ and $\tau = \begin{pmatrix} 0 & 1 \\ 1 & 0 \end{pmatrix}$ describe, in additive notation, how a generator for Z acts on A. Observe that the matrices σ and τ are conjugate over the rationals Q but not over the integers. It is easy to see that G contains an isomorphic copy of H of index 2 and that H contains a copy of G of index 2. Furthermore, $G \not\simeq H$ since G/G' has torsion but H/H' does not.

THEOREM 11 [9]. *Let G be a torsion-free polycyclic-by-finite group and set $D = K(G)$. Then linear groups over D enjoy similar properties to those over fields. For example*

(i) *Any periodic subgroup of $\mathrm{GL}_n(D)$ is locally finite (analog of Burnside's theorem).*

(ii) *If $\operatorname{char} K = 0$, then any periodic subgroup of $\mathrm{GL}_n(D)$ is abelian-by-finite (analog of Jordan's theorem).*

(iii) *If G is nilpotent, then any noncentral normal subgroup of $\mathrm{GL}_n(D)$ contains a noncyclic free group (analog of Tits' theorem).*

Part (i) has been further refined by Wehrfritz.

THEOREM 12 [10]. *Let G be a torsion-free polycyclic-by-finite group. If G is poly-(infinite cyclic), then $K(G)$ is a universal field of fractions of $K[G]$ (that is, a version of the Extension Theorem for Places holds). If G is not poly-(infinite cyclic), then this is not necessarily true.*

For a counterexample, take our original nasty group
$$H = \langle x, y \mid x^{-1} y^2 x = y^{-2}, y^{-1} x^2 y = x^{-2} \rangle$$
with $\operatorname{char} K \neq 2$ and let $\alpha \colon K[H] \to K$ be the augmentation map. If α extends to a specialization, then there exists an intermediate ring R with $K[H] \subseteq R \subseteq K(H)$ and an extended map $\alpha \colon R \to K$ such that $\operatorname{Ker} \alpha = JR$. Then R is local and all elements of $K[H]$ not in the augmentation ideal are invertible in R. Now compute.

In addition, Stafford has shown that for $D = K(G)$
$$\operatorname{Kdim} D^{\mathrm{op}} \otimes D = \operatorname{gl\,dim} D^{\mathrm{op}} \otimes D = h(G)$$
where the latter is the Hirsch number of G. Now let G be nilpotent. Then Makar-Limanov proved that $K(G)$ contains a noncommutative free subalgebra if G is nonabelian and Lorenz has studied the GK-transcendence degree of $K(G)$.

(3) Finally we discuss the Goldie rank problem. Here we assume that G is polycyclic-by-finite with $\Delta^+(G) = 1$ so $K[G]$ is prime. If N is a finite subgroup of G, then $|N| \mid \mathrm{u.d.}\, K[G]$. Conversely, if $|G : H| < \infty$ and $K[H]$ is a domain, then $\mathrm{u.d.}\, K[G] \mid |G : H|$. Thus

LEMMA 13. *In the above context* l.c.m.$_N |N| \mid$ u.d. $K[G]$ *and* u.d. $K[G] \mid$ g.c.d.$_H |G:H|$.

The conjecture is that u.d. $K[G] = $ l.c.m.$_N |N|$. In his work on this, Rosset formalized the argument of [5] as follows. Let H be any poly-(infinite cyclic) subgroup of G of finite index. If M is a finitely generated $K[G]$-module, let

$$0 \to P_n \to P_{n-1} \to \cdots \to P_0 \to M_H \to 0$$

be a projective resolution for the restricted module M_H. Since each projective $K[H]$-module P_i is stably free, we can define the Euler characteristic of M by

$$\chi(M) = |G:H|^{-1} \sum_i (-1)^i \operatorname{rank} P_i.$$

It is easy to see that this depends only upon M and not on the choice of H or the resolution. Furthermore

LEMMA 14. *With the above notation we have*
(i) *If M is a nonsingular $K[G]$-module, then $\chi(M) = $ u.d. $M/$ u.d. $K[G]$.*
(ii) *u.d. $K[G]$ is the smallest positive integer m with $m\chi(M) \in Z$ for all nonsingular modules M.*

Part (ii) follows since there exists a suitable module M with u.d. $M = 1$. The trace argument of [2] and [5] amounts to showing in the torsion-free case that if P is a projective $K[G]$-module then $|G:H|$ divides rank P_H. Hence $\chi(M)$ is always an integer. A nice application of the above is

PROPOSITION 15 [11]. *Let G be polycyclic-by-finite with $\Delta^+(G) = 1$. Then u.d. $Q[G]$ divides u.d. $GF(p)[G]$ for all primes p.*

PROOF. Write $F = GF(p)$. Let M be a uniform right ideal of $Q[G]$ and set $U = M \cap Z[G]$ so that $Q \otimes_Z U = M$. Since U is Z-pure in $Z[G]$ we see that $\overline{U} = F \otimes_Z U$ is contained in $F \otimes Z[G] = F[G]$ and thus $\overline{U}$ is a nonsingular $F[G]$-module.

Now let

$$0 \to P_n \to P_{n-1} \to \cdots \to P_0 \to U_H \to 0$$

be an appropriate projective resolution of $Z[H]$-modules. Since these are all free Z-modules, the sequence splits as Z-modules and tensoring with Q or F over Z is exact. Thus

$$0 \to Q \otimes P_n \to Q \otimes P_{n-1} \to \cdots \to Q \otimes P_0 \to M \to 0,$$
$$0 \to F \otimes P_n \to F \otimes P_{n-1} \to \cdots \to F \otimes P_0 \to \overline{U} \to 0.$$

are projective resolutions of $Q[H]$- and $F[H]$-modules respectively. Since

$$\operatorname{rank} Q \otimes P_i = \operatorname{rank} F \otimes P_i$$

we have $\chi(\overline{U}) = \chi(M)$ and thus, by Lemma 14,

$$\text{u.d.} \, \overline{U}/\text{u.d.} \, F[G] = 1/\text{u.d.} \, Q[G].$$

There does not appear to be a direct nonhomological proof of the above. We remark that the conjecture is known for G supersolvable, (free abelian)-by-cyclic, or poly-(infinite cyclic)-by-(elementary abelian p) in characteristic p.

References

1. K. A. Brown, *On zero divisors in group rings*, Bull. London Math. Soc. **8** (1976), 251–256.
2. G. H. Cliff, *Zero divisors and idempotents in group rings*, Canad. J. Math. **32** (1980), 596–602.
3. P. M. Cohn, *On the free product of associative rings*. III, J. Algebra **8** (1968), 376–383.
4. D. R. Farkas, A. H. Schofield, R. L. Snider and J. T. Stafford, *The isomorphism question for division rings of group rings*, Proc. Amer. Math. Soc. **85** (1982), 327–330.
5. D. R. Farkas and R. L. Snider, K_0 *and Noetherian group rings*, J. Algebra **42** (1976), 192–198.
6. F. T. Farrell and W. C. Hsiang, *The Whitehead group of poly-(finite or cyclic) groups*, J. London Math. Soc. (2) **24** (1981), 308–324.
7. E. Formanek, *Idempotents in Noetherian group rings*, Canad. J. Math. **15** (1973), 366–369.
8. ____, *The zero divisor question for supersolvable groups*, Bull. Austral. Math. Soc. **9** (1973), 67–71.
9. A. I. Lichtman, *On matrix rings and linear groups over a field of fractions of enveloping algebras and group rings* I, J. Algebra **88** (1984), 1–37.
10. D. S. Passman, *Universal fields of fractions for polycyclic group algebras*, Glasgow Math. J. **23** (1982), 103–113.
11. S. Rosset, *Miscellaneous results on the Goldie rank conjecture* (preprint).
12. J. P. Serre, *Cohomologie des groupes discrets*, Prospects in Mathematics (Proc. Sympos., Princeton Univ., 1970), Ann. of Math. Studies, No. 70, Princeton Univ. Press, Princeton, N. J., 1971, pp. 77–169.
13. R. Walker, *Local rings and normalizing sets of elements*, Proc. London Math. Soc. (3) **24** (1972), 27–45.

4. Polycyclic Group Rings

I recently wrote a survey [5] on group rings of polycyclic-by-finite groups. Since I do not wish to merely repeat parts of that here, I will stress different aspects. The subject has been flourishing because of its connection with Noetherian ring theory. Nevertheless, there has been a strong influence from the study of division rings and from group theory. We consider some examples.

Let D be a divison ring with center K. Suppose G is a multiplicative subgroup of $D\backslash 0$ and assume that G and K generate D as a division ring. The latter implies that $Z(G) = G \cap K$. The following result of Zalesskii indicates how G sits in D, at least for certain special G.

LEMMA 1. *Let D, K and G be as above and assume that G is torsion-free nilpotent. If T is a transversal for $Z(G)$ in G, then the elements of $T \subseteq D$ are linearly independent over K.*

PROOF. Let G_n denote the nth term of the upper central series of G and let R_n be the K-linear span of G_n in D. Thus $R_1 = K$. For $n \geq 1$ let T_{n+1} be a transversal for G_n in G_{n+1} with $1 \in T_{n+1}$. It suffices to show that T_{n+1} is free over R_n for all $n \geq 1$.

Suppose $\sum_0^t \alpha_i = 0$ with $\alpha_i \in R_n x_i$ and $x_i \in T_{n+1}$. We can assume that t is minimal, all $\alpha_i \neq 0$ so $t \geq 1$ and that $x_0 = 1$. If $g \in G$ we have $\sum_0^t \alpha_i{}^g = 0$ so

$$\sum_{i=0}^t (\alpha_0 \alpha_i{}^g - \alpha_i \alpha_0{}^g) = 0.$$

Now it is easy to see that the ith term here is contained in $R_n x_i$ and that the $i = 0$ term is missing. Thus the minimality of t implies that all terms vanish. Write $\alpha_0 = \alpha$, $\alpha_1 = \beta x$ with $\alpha, \beta \in R_n \backslash 0$ and $x \in G_{n+1} \backslash G_n$. Then we obtain the identity

$$\alpha(\beta x)^g = (\beta x)\alpha^g \qquad \forall g \in G.$$

If $n = 1$, then $R_1 = K = Z(D)$. Thus the above yields $x^g = x$, $\forall g \in G$, so $x \in G_1$, a contradiction. Finally let $n > 1$ and rewrite the identity as

$$\alpha \beta^g (x^g x^{-1}) = \beta \alpha^{g x^{-1}} \qquad \forall g \in G.$$

Since $x^g x^{-1} \in G_n$, this is an equation in R_n. By induction, $R_n = R_{n-1} T_n$ with T_n free over R_{n-1}. Thus each element of R_n has a well defined support in

$T_n = G_n/G_{n-1}$. Since this is a central factor of G, we have $\operatorname{Supp} \beta^g = \operatorname{Supp} \beta$ and similarly for α. Also multiplication is in D so products do not vanish. Thus for all $g \in G$
$$x^g x^{-1} = b_1^{-1} a_1^{-1} b_2 a_2 \quad \bmod G_{n-1}$$
with $a_i \in \operatorname{Supp} \alpha$, $b_i \in \operatorname{Supp} \beta$. In other words, $x^g x^{-1}$ takes on only finitely many values $\bmod G_{n-1}$ so $xG_{n-1} \in \Delta(G/G_{n-1}) = G_n/G_{n-1}$ since G is torsion-free nilpotent. This yields $x \in G_n$, a contradiction.

We remark that for $n \geq 1$, $R_n = R_{n-1} * (G_n/G_{n-1})$ is actually a crossed product (which will be defined later on). Now the embedding of G in D gives rise to a homomorphism $K[G] \to D$ whose kernel P is obviously a prime ideal. Note that $1 - g \in P$ if and only if $g = 1$. Furthermore, the above lemma says that $P = (P \cap K[G_1]) \cdot K[G]$. This was the first indication that primes might have this property for more general groups. Indeed it asserts that $G_1 = Z(G)$ controls P.

Let G be a polycyclic-by-finite group which is not necessarily nilpotent. Then $Z(G)$ may be trivial, but G can still have normal abelian subgroups of interest. In fact, we will consider what the *minimal* such might look like. Let $1 \neq A \triangleleft G$ with A free abelian and assume that no nonidentity subgroup of A of smaller rank is normal in G. Then in additive notation, G acts irreducibly on the finite-dimensional rational vector space $Q \otimes A^+$. We say that G acts rationally irreducibly on A. It is quite possible that A becomes reducible for a subgroup of finite index in G. Thus, at the expense of dropping down to a subgroup of G of finite index, we get a more interesting concept.

Let $1 \neq A$ be a finitely generated free abelian group and let G be an arbitrary group acting on A. We say that A is a plinth for G if G and all its subgroups of finite index act rationally irreducibly on A. It was Zalesskii who first saw the importance of plinths. Note that G also acts on $K[A]$ and permutes its ideals. Zalesskii believed that the G-invariant primes of $K[A]$ were necessarily scarce. In fact, he conjectured the following beautiful result of Bergman.

THEOREM 2 [1]. *Let A be a plinth for G. If P is a G-invariant prime ideal of $K[A]$, then $P = 0$ or $K[A]/P$ is finite-dimensional over K.*

PROOF (FLAVOR). Assume that $K = Q$, the rationals, and that $A = \langle x, y \rangle$. Then $Q[A]/P$ is a commutative domain containing Q, and the images $\bar{x}$, $\bar{y}$. If $\bar{x}$ and $\bar{y}$ are both algebraic over Q, then $Q[A]/P = \overline{Q[A]}$ is a finite field extension of the rationals and we are done. Thus we may assume that $\bar{x}$ is transcendental. We can then embed $\overline{Q[A]}$ into the complex numbers C in many different ways. In particular we can insist that $|\bar{x}| \neq 1$.

Suppose $\alpha \in P$, $\alpha \neq 0$ and write $\alpha = \sum_{t \in A} k_t t$. Then $0 = \sum k_t \bar{t} \in C$. Let $u \neq v$ be elements of $\operatorname{Supp} \alpha$ with $|\bar{u}|$ maximal and $|\bar{v}|$ second maximal. Then $k_u \bar{u} = -\sum_{t \neq u} k_t \bar{t}$ implies that
$$|\bar{u}| \leq |k_u|^{-1} \sum_t |k_t| \cdot |\bar{v}|$$

and hence $1 \le |\overline{uv^{-1}}| \le N$ where N depends only on the coefficients of α and not on the choice of u or v.

Now P is G-invariant, so if $g \in G$ then $\alpha^g = \sum k_t t^g \in P$. We conclude as above that $1 \le |\overline{(uv^{-1})^g}| \le N$ for the same N but for perhaps different $u, v \in \operatorname{Supp} \alpha$. But there are only finitely many possible pairs to deal with, so for simplicity let us assume that we always get $uv^{-1} = z$. Then for all $g \in G$, $1 \le |\overline{z^g}| \le N$. For this $z \ne 1$ write $z^g = x^{a(g)} y^{b(g)}$ so that $|\overline{z^g}| = |\overline{x}|^{a(g)} |\overline{y}|^{b(g)}$. Taking logs then yields

$$0 \le a(g) \ln |\overline{x}| + b(g) \ln |\overline{y}| \le \ln N.$$

In the plane, let ν denote the vector $\nu = \ln |\overline{x}| i + \ln |\overline{y}| j$ so that $\nu \ne 0$ since $|\overline{x}| \ne 1$. Then the above inequality forces the vector $a(g)i + b(g)j$ to belong to an infinite strip perpendicular to ν and of finite width. (See the figure below for an example.)

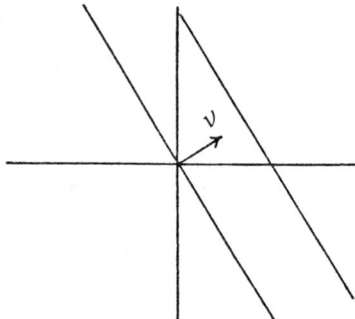

Let $M \colon G \to \operatorname{GL}_2(Z)$ denote the matrix representation of G on A. Then for all $g \in G$, $[a(1), b(1)] M(g) = [a(g), b(g)]$. Now in general, for $g \ne 1$, $M(g)$ will have a dominant eigenvalue and hence by applying $M(g^n)$ for $n \to \infty, -\infty$ we see that $M(g)$ must have an eigenvector perpendicular to ν. Since this is true for all $g \in G$, G does not act irreducibly on A.

Alternately we can vary the embedding of $\overline{Q[A]}$ into C getting other choices for ν. Bergman extended this idea to a careful characteristic 0 proof. Then he eliminated all the analysis by using non-Archimedean valuations which apply in all characteristics.

We remark that if $P \ne 0$, then G acts on the finite-dimensional field extension $K[A]/P$ fixing K and thus G must act like a finite group. In other words, G has a subgroup H of finite index which acts trivially on $\overline{A}$, the image of A. If $\overline{A} \simeq A$, then H acts trivially on A and the plinth assumption implies that $\operatorname{rank} A = 1$. Otherwise $\overline{A} \not\simeq A$. So $\overline{A}$ is finite and P is the complete inverse image of a G-invariant prime of $K[\overline{A}]$.

This leads to the following definition. Let $I \triangleleft K[G]$ and observe that $G \subseteq K[G]$ maps into the group of units of $K[G]/I$. We say that I is faithful if G embeds in $K[G]/I$. I is almost faithful if the kernel of the homomorphism is finite. The key part of the next result is the easing up of the assumptions on the group action. The more general A is not difficult to consider.

THEOREM 3 [8]. *Let A be a finitely generated f.c.-group, let G act on A and set $D = D_A(G)$, the almost centralizer of G in A. Then $D \triangleleft A$, and if P is an almost faithful G-invariant prime of $K[A]$ we have $P = (P \cap K[D]) \cdot K[A]$ so that D controls P.*

Let us move on to a second influx of ideas, namely the group theoretic work of Ph. Hall. Hall was interested in extending known properties of polycyclic-by-finite groups to more general families. In particular he studied finitely generated abelian-by-(polycyclic-by-finite) groups. For our purposes they look like $M \rtimes G$ where M is abelian, G is polycyclic-by-finite, and M^+ is a finitely generated $Z[G]$-module. Obviously, the study of such groups is equivalent to the study of $Z[G]$-modules.

For example, polycyclic-by-finite groups are residually finite. Hall asked about the larger class. Let us see what is involved here. Let M be an irreducible $Z[G]$-module. If $M \rtimes G$ is residually finite then it has a normal subgroup N of finite index with $N \not\supseteq M$. But then $N \cap M$ corresponds to a proper submodule so $N \cap M = 1$ and M embeds in the finite quotient $(M \rtimes G)/N$. Thus, at the very least we must show that if M is an irreducible $Z[G]$-module, then M is finite. In particular this says that for any such M there exists a rational prime p with $Mp = 0$, and furthermore, all irreducible $GF(p)[G]$-modules are finite. The latter statement is a natural generalization of the Hilbert Nullstellensatz. Indeed if G is a finitely generated abelian group, then one aspect of the Nullstellensatz asserts that all irreducible $K[G]$-modules are finite-dimensional over K.

To deal with this problem, Hall discovered a new technique for proving the Nullstellensatz. This method is now known as generic flatness.

THEOREM 4 [2]. *Let G be a polycyclic-by-finite group, let A be a torsion-free central subgroup of G and let R be a commutative Noetherian domain. Suppose M is a finitely generated $R[G]$-module. Then there exists a free $R[A]$-submodule M_0 of M and $0 \neq \alpha \in R[A]$ such that every element of M/M_0 is annihilated by some power of α.*

In particular, if we localize at the element α given above, then $M_\alpha = (M_0)_\alpha$ is a free $R[A]_\alpha$-module.

COROLLARY 5. *With the above notation, if M is irreducible and if $R[A]$ is a principal ideal domain with infinitely many primes, then $R[A]$ does not act faithfully on M.*

PROOF. Assume that $R[A]$ acts faithfully. Since A is central in G this implies that $R[A]$ embeds in the division ring $D = \text{End}_{R[G]} M$. Let $m \in M \setminus 0$ and let F be the field of fractions of $R[A]$. Then F embeds in D and $M \supseteq Dm \supseteq Fm \simeq F$ as $R[A]$-modules. Since a submodule of a free $R[A]$-module is free, there exists F_0 a free $R[A]$-submodule of F and $0 \neq \alpha \in R[A]$ such that every element of

F/F_0 is annihilated by a power of α. Of course we must have $F_0 = \beta R[A]$ for some $\beta \in F$, so it follows that every element of F is contained in $\beta \alpha^{-n} R[A]$ for some n. In particular, only finitely many primes can occur as denominators in F.

Since Z and $GF(p)[\langle x \rangle]$ are principal ideal domains we get

COROLLARY 6 [2]. *Let G be a polycyclic-by-finite group and let M be an irreducible $Z[G]$-module. Then there exists a prime p with $Mp = 0$. Furthermore, if G is nilpotent, then M is finite.*

The problem was finally solved in the affirmative by Roseblade and Jategoankar. A key intermediate result is the following Nullstellensatz with operators.

THEOREM 7 [4, 6]. *Let A be a finitely generated free abelian group which is a plinth for the polycyclic-by-finite group G. Let $\alpha \in K[A]$ and suppose that for every maximal ideal T of $K[A]$ with only finitely many G-conjugates we have $\alpha^g \in T$ for some $g \in G$. Then $\alpha = 0$.*

A second group theoretic question concerns the nilpotence of the Frattini subgroup. Recall that $\Phi(G)$ in the intersection of all maximal subgroups of G. It follows that if $G \to \overline{G}$ is an epimorphism, then $\Phi(G)$ maps into $\Phi(\overline{G})$. If G is polycyclic-by-finite, then a result of Hirsch asserts that $\Phi(G)$ is nilpotent. How about $W = M \rtimes G$? Let us see what is involved here.

Let H be the image of $\Phi(W)$ in $W/M = G$. Thus $H \subseteq \Phi(G)$ is a normal nilpotent subgroup of G. Let T be a maximal submodule of M. Then M/T is finite so $W/T = (M/T) \rtimes G$ is polycyclic-by-finite. Since $\Phi(W)$ maps into the nilpotent group $\Phi(W/T)$, we deduce that $\Phi(W)$ centralizes M/T and hence H centralizes M/T for all such T. Note that $Z[H]$ acts on M and if I is its augmentation ideal, then I annihilates each M/T.

To see what this means, let us consider an example. Suppose that $M = Z[G]$ is the regular $Z[G]$-module. Then each T is a maximal right ideal so, since I annihilates M/T, we have $I \subseteq T$. In other words, $I \subseteq JZ[G]$. The best we can hope for here, in analogy with the Nullstellensatz, is that I is nilpotent.

Returning to the general problem, we expect to conclude that I is nilpotent in its action on M. This was shown by Hall for G a nilpotent group and then

THEOREM 8 [7]. *Let G be a polycyclic-by-finite group, M a finitely generated $Z[G]$-module, and let H be a normal nilpotent subgroup of G. If the augmentation ideal I of $Z[H]$ annihilates all quotients M/T with T a maximal $Z[G]$-submodule, then $MI^n = 0$ for some n.*

COROLLARY 9. *If W is a finitely generated abelian-by-(polycyclic-by-finite) group, then $\Phi(W)$ is nilpotent.*

PROOF. Say $W = M \rtimes G$ and use the above notation. We know that H is a normal nilpotent subgroup whose augmentation ideal I kills all M/T. Therefore,

$MI^n = 0$ for some n. In multiplicative notation this says that the n-fold commutator $[M, H, H, \ldots, H] = 1$. Since H is nilpotent it follows easily that $M \rtimes H$ is nilpotent and hence so is $\Phi(W) \subseteq M \rtimes H$.

Another aspect of the Nullstellensatz was observed by Segal. An R-module M is poly-(residually simple) if M has a finite chain of submodules $0 = M_0 \subseteq \cdots \subseteq M_n = M$ with each quotient M_{i+1}/M_i residually simple (that is, has radical zero). Segal proved the marvelous

THEOREM 10 [9]. *Let G be a finitely generated nilpotent group. Then any finitely generated $Z[G]$-module is residually simple.*

PROOF (FOR G ABELIAN). Since M is a Noetherian $Z[G]$-module we can let M' be a maximal poly-(residually simple) submodule. We can now mod out by M'. The goal is to show that $M = 0$. If $M \neq 0$, then M contains a nonzero element m whose annihilator in $Z[G]$ is a prime ideal P. Thus $mZ[G] \simeq Z[G]/P$. But the Nullstellensatz implies that any prime ideal is an intersection of maximal ideals so $mZ[G]$ is residually simple. This contradicts the maximality of M'.

The noncommutative proof is orders of magnitude more difficult. It was then extended by Brookes to arbitrary polycyclic-by-finite groups.

A third input into the subject is the general theory of noncommutative Noetherian rings. If G is polycyclic-by-finite, then $K[G]$ is Noetherian. In fact, it is sometimes hard to find where group ring theory ends and Noetherian ring theory begins. We will not discuss additional motivation here except to say that prime ideals and prime rings are the basic building blocks of the Goldie theory. The best result in this part of group rings is a description of the primes of $K[G]$.

Let P be a prime of $K[G]$. We say that P is standard if:

1. $P = (P \cap K[\Delta]) \cdot K[G]$ so that $\Delta(G)$ controls P.
2. $P \cap K[\Delta] = Q_1 \cap Q_2 \cap \cdots \cap Q_n$ a finite intersection of G-conjugate almost faithful primes of $K[\Delta]$.

Note that Δ is center-by-finite and finite-by-abelian so the individual Q_i are well understood. Also, $G/\mathbb{C}_G(\Delta)$ is finite so G acts as a finite group on Δ. We remark that if P is standard then it is almost faithful.

More generally we say that P is image standard if there exists $N \triangleleft G$ with P the complete inverse image in $K[G]$ of a standard prime of $K[G/N]$. Since the natural map $K[G] \to K[G/N]$ is so simple, the structure of image standard primes is also well understood.

If G is a torsion-free, finitely generated nilpotent group, then Zalesskii showed that all faithful primes of $K[G]$ are standard. More generally, there exists a property known as being orbitally sound which is satisfied by nilpotent groups and for which we have

THEOREM 11 [8]. *Let G be a polycyclic-by-finite group.*

(i) *There exists a unique largest subgroup G_0 of finite index in G which is orbitally sound.*

(ii) *If G is orbitally sound then all prime ideals of $K[G]$ are image standard.*

It follows from later work that the converse of (ii) also holds. That is, if all prime ideals of $K[G]$ are image standard, then G is orbitally sound. We will discuss definitions and results for arbitrary polycyclic-by-finite groups later on. However, for many problems the finite index $|G : G_0|$ is easily finessed. In particular, there exist applications to describing primitive ideals, computing prime and primitive lengths, and describing division rings generated by polycyclic-by-finite groups.

We close with two problems of interest. The first one asks whether $K[G]$ is a catenary ring; that is, for all primes $P \subseteq Q \neq K[G]$, are all saturated chains of primes joining P and Q of equal length? This is known to be true for G orbitally sound, but the finite index problem has proved to be difficult to surmount. The second is to study the structure of the injective $K[G]$-modules. Since $K[G]$ is Noetherian, each injective module is uniquely a direct sum of uniform injectives. Thus we are interested in the injective hulls $E(X)$ for X a uniform $K[G]$-module. In particular we want to study essential extensions of uniform modules. For example, let P, Q be prime ideals of $K[G]$. Then we say that there exists a link $P \rightsquigarrow Q$ if and only if (roughly) there exists a short exact sequence $0 \to X \to U \to Y \to 0$ where U is uniform, $X \neq 0$ is a uniform right ideal of $K[G]/P$, and $Y \neq 0$ is a uniform right ideal of $K[G]/Q$. Such links have been studied by Brown and Warfield.

References

1. G. M. Bergman, *The logarithmic limit-set of an algebraic variety*, Trans. Amer. Math. Soc. **157** (1971), 459–469.
2. P. Hall, *On the finiteness of certain soluble groups*, Proc. London Math. Soc. (3) **9** (1959), 595–622.
3. ____, *The Frattini subgroups of finitely generated groups*, Proc. London Math. Soc. (3) **11** (1961), 327–352.
4. D. L. Harper, *Primitivity in representations of polycyclic groups*, Math. Proc. Cambridge Philos. Soc. **88** (1980), 15–31.
5. D. S. Passman, *Group rings of polycyclic groups*, Group Theory: Essays for Philip Hall, Academic Press, London, 1984, pp. 207–256.
6. J. E. Roseblade, *Group rings of polycyclic groups*, J. Pure Appl. Algebra **3** (1973), 307–328.
7. ____, *The Frattini subgroup in infinite soluble groups*, Three Lectures on Polycyclic Groups, Queen Mary College Mathematics Notes, Queen Mary College, London, 1973.
8. ____, *Prime ideals in group rings of polycyclic groups*, Proc. London Math. Soc. (3) **36** (1978), 385–447; Corrigenda, **38** (1979), 216–218.
9. D. Segal, *On the residual simplicity of certain modules*, Proc. London Math. Soc. (3) **34** (1977), 327–353.

5. Crossed Products of Finite Groups

Let R be a ring and let G be a group. To start with, a crossed product $R*G$ is a generalized group ring. It has as an R-basis the set $\overline{G}$ which is a copy of G, so that each element of $R*G$ is uniquely a finite sum $\sum_{x \in G} r_x \overline{x}$ with $r_x \in R$. Addition is as expected, but multiplication has two new wrinkles, a twisting and an action. Specifically, for $x, y \in G$, we have

(twisting) $$\overline{x}\,\overline{y} = t(x,y)\overline{xy}$$

where $t \colon G \times G \to U = U(R)$, the group of units of R. Furthermore, for $x \in G$ and $r \in R$, we have

(action) $$r\overline{x} = \overline{x} r^{\overline{x}}$$

where $\overline{x} \in \operatorname{Aut} R$. The twisting and action are interrelated by conditions precisely equivalent to $R*G$ being associative. Note that we can and will assume that $\overline{1} = 1$. It follows that R is naturally embedded in the crossed product via $r \to r\overline{1}$. On the other hand, G is in general not contained in $R*G$. Nevertheless, each $\overline{x}$ is a unit in the ring, $\mathfrak{G} = \{u\overline{x} \mid u \in U,\ x \in G\}$ is the group of so-called trivial units of $R*G$ and $\mathfrak{G}/U \simeq G$.

We study crossed products because they occur naturally. We do not merely go around constructing them. For example

LEMMA 1. *Let $N \triangleleft H$. Then $K[H] = K[N] * (H/N)$.*

PROOF. Set $R = K[N]$ and $G = H/N$. For each $x \in G$ let $\overline{x} \in H$ be a fixed inverse image. Then $H = \bigcup_x N\overline{x}$ implies that

$$K[H] = \oplus \sum_x K[N]\overline{x} = \oplus \sum_x R\overline{x}$$

so $\overline{G}$ is an R-basis for $K[H]$.

Since $N \triangleleft H$, $\overline{x}^{-1}N\overline{x} = N$ so $\overline{x}^{-1}K[N]\overline{x} = K[N]$ and $\overline{x}$ induces a conjugation automorphism on R. In particular, if $x \in G$ and $r \in R$, then

$$r\overline{x} = \overline{x}(\overline{x}^{-1}r\overline{x}) = \overline{x}r^{\overline{x}}$$

and we see that an action occurs. Note that the action is trivial if N is central in H.

Next, for $x,y \in G$ we have $N\bar{x} \cdot N\bar{y} = N\overline{xy}$ so $\bar{x}\bar{y} = t(x,y)\overline{xy}$ for some $t(x,y) \in N \subseteq U(R)$. Furthermore, this twisting is trivial if we can choose a consistent set of coset representatives, that is, if $H = N \rtimes G$. Finally we observe that $K[H]$ is certainly associative.

Several remarks are in order. (1) Finite index problems occur frequently in the study of group rings. Namely, suppose we know information about $K[N]$ with N a normal subgroup of H of finite index. The goal is to lift this information to $K[H]$. Since $K[H] = K[N] * (H/N)$, the structure of crossed products of finite groups can sometimes help. We will offer a nice example of this. (2) Suppose $I \triangleleft K[H]$ is controlled by N so that $I = L \cdot K[H]$ with $L = I \cap K[N]$. It then follows easily that $K[H]/I = (K[N]/L) * (H/N)$. (3) Finally, the same argument shows that if we are given $R * H$ and $N \triangleleft H$, then $R * H \supseteq R * N$ and $R * H = (R * N) * (H/N)$. Thus we do not leave the family of crossed products.

Certain special cases of crossed products have their own names. If there is no action or twisting, then $R * G = R[G]$ is an ordinary group ring. If the action is trivial, then $R * G = R^t[G]$ is a twisted group ring. Finally, if the twisting is trivial, then $R * G = RG$ is a skew group ring. We frequently construct the latter.

LEMMA 2. *Let $G \to \text{Aut}\, R$ be a group homomorphism and define RG as above. Then this skew group ring is associative.*

Note that since the twisting is trivial in RG we have $\bar{x}\bar{y} = \overline{xy}$. Thus we can drop the overbars here and assume that $RG \supseteq G$. The skew group ring is a useful tool in the Galois theory of rings. We will discuss this later on.

Historically, crossed products arose in the study of division rings. Let K be a field and let D be a division algebra finite-dimensional over its center K. If F is a maximal subfield of D, then $\dim_K D = (\dim_K F)^2$. Suppose that F/K is normal, although this is not always true. If $x \in \text{Gal}(F/K) = G$, then the Skolem-Noether theorem implies that there exists $\bar{x} \in D\backslash 0$ with $\bar{x}^{-1}f\bar{x} = f^x$ for all $f \in F$. Furthermore, $\bar{x}\bar{y}$ and $\overline{xy}$ agree in their action on F so $\overline{xy}\bar{x}^{-1}\bar{y}^{-1} \in C_D(F) = F$. Once we show that the elements $\bar{x}$ for $x \in G$ are linearly independent over F, then we conclude by computing dimensions that $D = \oplus \sum_{x \in G} F\bar{x} = F * G$.

More generally we say that A is central simple over K, if A is a finite-dimensional simple algebra over its center K. Thus $A = M_n(D)$ for some n and division ring D with $Z(D) = K$. Two such algebras A and B are equivalent if they have the same D. The equivalence classes then form a group under tensor product $\otimes_K$, the Brauer group. Now given A, one can show that there exists $B \sim A$ with $B = F * G$. But $F * G$ is determined by the twisting function $t: G \times G \to F$, a 2-cocycle. Thus, in this way we obtain the homological characterization of the Brauer group as the 2nd cohomology group.

We now begin to consider the ring theoretic properties of $R * G$ with G finite. We start with the essential version of Maschke's theorem (see [**7**]).

THEOREM 3. *Given $R*G$ with G finite. Let $W \subseteq V$ be $R*G$-modules with no $|G|$-torsion. Then $W \operatorname{ess}_R V$ if and only if $W \operatorname{ess}_{R*G} V$.*

PROOF. We need only show that $W \operatorname{ess}_{R*G} V$ implies that $W \operatorname{ess}_R V$ since the other implication is obvious.

Case 1. Suppose $V = W \oplus U$ where U is a complementary R-submodule. Then for all $x \in G$, $V = W\bar{x} \oplus U\bar{x} = W \oplus U\bar{x}$ and we let $\pi_x \colon V \to W$ be the R-homomorphism determined by this decomposition. Note that if $v = w + u\bar{x}$, then $v\bar{y} = w\bar{y} + u\bar{x}\,\bar{y}$, so clearly

$$\pi_{xy}(v\bar{y}) = w\bar{y} = \pi_x(v)\bar{y}.$$

It follows that $\pi = \sum_x \pi_x$ is an $R*G$-homomorphism from V to W since

$$\pi(v\bar{y}) = \sum_x \pi_x(v\bar{y}) = \sum_x \pi_{xy}(v\bar{y})$$
$$= \sum_x \pi_x(v)\bar{y} = \pi(v)\bar{y}.$$

Furthermore, we have $\pi(w) = |G|w$ so $\operatorname{Ker} \pi \cap W = 0$ since V has no $|G|$-torsion, and hence, $\operatorname{Ker} \pi = 0$ since $W \operatorname{ess}_{R*G} V$. Finally, let $u \in U$ and set $w = \pi(u)$. Then $|G|u - w \in \operatorname{Ker} \pi$ so $|G|u \in W \cap U = 0$ and $u = 0$.

Case 2. Now for the general case. Choose $U_R \subseteq V$ maximal with $U_R \cap W = 0$. Then $W \oplus U \operatorname{ess}_R V$ and set $E = \bigcap_x (W \oplus U)\bar{x}$. It follows that $E \operatorname{ess}_R V$ and that E is an $R*G$-submodule. Furthermore, $W \subseteq E \subseteq W \oplus U$ so $E = W \oplus (U \cap E)$. By Case 1, $W = E$ so $W \operatorname{ess}_R V$.

THEOREM 4 [2]. *Let R be a semiprime ring with no $|G|$-torsion. Then $R*G$ is semiprime.*

PROOF. Let $N \triangleleft R*G$ with $N^2 = 0$. If $L = l_{R*G}(N)$, then $L \triangleleft R*G$ and $L \operatorname{ess}_{R*G} R*G$ as right ideals. Maschke's theorem now implies that $L \operatorname{ess}_R R*G$ so $(L \cap R) \operatorname{ess}_R R$. Since R is semiprime, we conclude that $r_R(L \cap R) = 0$ and then $N \subseteq r_{R*G}(L \cap R) = 0$, by the freeness of $R*G$ over R.

The proof given above is from [7] but the original techniques are still needed for the following generalization.

THEOREM 5 [8]. *Given $R*G$ with G finite.*
(i) *If $R*G$ is semiprime and $H \subseteq G$, then $R*H$ is semiprime.*
(ii) *Assume that $R*P$ is semiprime for $P = 1$ and all elementary abelian p-subgroups $P \subseteq G$ such that R has p-torsion. Then $R*G$ is semiprime.*

Now let us study the prime ideals in $R*G$ with G finite. Note that G permutes the ideals of R by conjugation. If $A \triangleleft R*G$, then $A \cap R$ is a G-invariant ideal of R. Conversely, if I is a G-invariant ideal of R, then $I*G = I(R*G)$ is an ideal of R with $(I*G) \cap R = I$. Moreover, $(R*G)/(I*G) = (R/I)*G$.

To study the prime ideals P of $R*G$ we might as well mod out by $(P \cap R)*G$ and assume that $P \cap R = 0$. This forces R to be G-prime, a condition somewhat weaker than being prime. Indeed for G finite it means that there exists a prime ideal Q of R with $\bigcap_x Q^{\bar{x}} = 0$. Note that $\{Q^{\bar{x}} | x \in G\}$ is the set of minimal primes of R and hence is uniquely determined by R. There are two cases to consider according to whether R is prime or not. We start with the latter situation.

Let H be a subgroup of G and let $I \triangleleft R * H$. Then $I(R * G)$ is a right ideal of $R * G$ and we denote by I^G the unique largest two-sided ideal it contains. In other words, the induced ideal I^G is given by

$$I^G = \mathrm{Id}(I(R*G)) = \bigcap_{x \in G} (I(R*G))^{\bar{x}}.$$

If J is a second ideal of $R * H$, then $I^G J^G \triangleleft R * G$ and

$$I^G J^G \subseteq I(R*G) J^G \subseteq I J^G \subseteq IJ(R*G).$$

Thus, induction satisfies the submultiplicative formula $I^G J^G \subseteq (IJ)^G$. Note that $H \triangleleft G$ implies that

$$I^G = \left(\bigcap_{x \in G} I^{\bar{x}} \right) (R*G)$$

and H controls I^G. The first main result on primes is

THEOREM 6 [5]. *Given $R*G$ with G finite and suppose Q is a prime ideal of R with $\bigcap_{x \in G} Q^{\bar{x}} = 0$. Let H be the stabilizer of Q in G. Then the map $T \to T^G$ yields a one-to-one correspondence between the primes T of $R*H$ with $T \cap R = Q$ and the prime ideals P of $R*G$ with $P \cap R = 0$.*

Note that there is an obvious one-to-one correspondence between the primes T of $R*H$ with $T \cap R = Q$ and the prime ideals $\overline{T}$ of $(R*H)/(Q*H) = (R/Q)*H$ with $\overline{T} \cap (R/Q) = 0$. This, therefore, reduces considerations to the prime case.

Now suppose that R is prime and consider all R-module homomorphisms $f: {}_R A \to {}_R R$ where A runs over all nonzero two-sided ideals of R. We say that $f \sim g$ if and only if f and g agree on their common domain and we let $\hat{f}$ denote the equivalence class of f. Then $Q_l(R)$, the set of all such equivalence classes, becomes a ring under function addition and composition. Furthermore, R embeds in $Q_l(R)$ via $r \to$ right multiplication by r. $Q_l(R)$ is called the (left) Martindale ring of quotients of R.

We will discuss $S = Q_l$ in more detail later on. For now it suffices to know that $C = Z(S)$ is a field called the extended centroid of R. Also, any automorphism σ of R extends uniquely to one of S. We say that σ is X-inner if it becomes inner on S.

Given $R * G$, there exists a unique extension to a crossed product $S * G$. We let $G_{\mathrm{inn}} = \{x \in G | \bar{x} \text{ is } X\text{-inner on } R\}$. Then $G_{\mathrm{inn}} \triangleleft G$ and the second main result on primes is

THEOREM 7 [5]. *Let $R * G$ be a crossed product with G finite and R prime. Set $S = Q_l(R)$ and let $E = C_{S*G}(S)$.*

(i) $E = C^t[G_{\text{inn}}]$ is a twisted group algebra of G_{inn} over the field C, the extended centroid of R.

(ii) Conjugation by each $\bar{x} \in R*G \subseteq S*G$ yields an action of G on E.

(iii) Their exists a one-to-one correspondence between the prime ideals P of $R*G$ with $P \cap R = 0$ and the G-orbits of primes of E.

To be precise, the G-orbit $\{T^{\bar{x}}\}$ of primes of E corresponds to

$$P = \left(\bigcap_{x \in G} T^{\bar{x}}\right) \cdot (S*G) \cap (R*G).$$

By combining the above two results we see that for R a G-prime ring, the primes P of $R*G$ with $P \cap R = 0$ correspond to the H-orbits of primes of $E = C^t[H_{\text{inn}}]$ where C is the extended centroid of R/Q. Since E is a finite-dimensional C-algebra, this implies that there are only finitely many primes P and indeed we have

COROLLARY 8. *Given $R*G$ with G finite and R a G-prime ring.*

(i) *If P is a prime ideal of $R*G$, then P is a minimal prime if and only if $P \cap R = 0$.*

(ii) *There exist only finitely many minimal primes $P_1, P_2, \ldots, P_n$ and $n \leq |G|$.*

(iii) *If $J = P_1 \cap P_2 \cap \cdots \cap P_n$, then J is the unique largest nilpotent ideal of $R*G$ and $J^{|G|} = 0$.*

This result is now a special case of properties of finite normalizing extensions. But more information is available here. For example, in the notation of Theorem 6, the number of primes P with $P \cap R = 0$ is at most equal to the number of conjugacy classes of H (not of G). Furthermore, if $\operatorname{char} C = p > 0$, then only the p-regular classes matter. In particular, if H_{inn} is a p-group, then there exists a unique prime P with $P \cap R = 0$.

We can also use these results to describe when $R*G$ is prime. It reduces to the twisted group algebra case which is still unsolved.

The above material was developed to study prime ideals in group rings of polycyclic-by-finite groups. For example, if G is such a group, then G has a normal subgroup G_0 of finite index which is orbitally sound, and hence, we essentially know the primes of $K[G_0]$. Furthermore, $K[G] = K[G_0] * (G/G_0)$ so the preceding results can apply. Indeed, Theorem 6 translates almost directly as follows. Let P be a prime ideal of $K[G]$ and let Q be a prime of $K[G_0]$ minimal over $P \cap K[G_0]$. If H is the stabilizer of Q in G, then $G \supseteq H \supseteq G_0$ and there exists a prime T of $R*(H/G_0) = K[H]$ with

$$P = T^G = \bigcap_{x \in G}(T \cdot K[G])^x.$$

It remains to consider T and, in an attempt to apply Theorem 7 directly, Lorenz and I computed X-inner automorphisms using Δ-methods. It turned out that yet another variant of the Δ-method was the missing ingredient, namely

PROPOSITION 9 [4]. *Let G be a polycyclic-by-finite group and let I be an ideal of $K[G]$ with $I = (I \cap K[\Delta]) \cdot K[G]$ and $I \cap K[\Delta] = Q_1 \cap Q_2 \cap \cdots \cap Q_n$, an intersection of almost faithful primes. If $A, B \triangleleft K[G]$ with $AB \subseteq I$, then $\theta(A)\theta(B) \subseteq \theta(I) = I \cap K[\Delta]$. In particular, I is a semiprime ideal and if all Q_i are G-conjugate, then I is a (standard) prime.*

This was sufficient to show that T above is image standard. Thus, to start with, if P is a prime ideal of $K[G]$, then $P = T^G$ for T an image standard prime of $K[H]$ with $|G : H| < \infty$. This sounds like a complete solution but it does leave a number of questions unanswered. Most notably we ask what are the possibilities for H and how unique is the situation. For this we need some additional definitions.

Let N be a subgroup of G. Then N is orbital if and only if it has a finite number of G-conjugates, that is, $|G : \mathsf{N}_G(N)| < \infty$. N is said to be an isolated orbital if it is orbital and for any larger orbital $N_1 > N$ we have $|N_1 : N| = \infty$. Finally, G is orbitally sound if all isolated orbitals are normal. Note that any finite orbital subgroup of G is contained in $\Delta^+(G)$ and that Δ^+ is an isolated orbital. The answer to the first question is that H can be taken to be the normalizer of an isolated orbital subgroup. Thus if G is orbitally sound, then $H = G$. Next, if $H = \mathsf{N}_G(N)$, then what is N? Again we need some more definitions.

If $I \triangleleft K[G]$, then we let $I^\dagger = \{x \in G | x - 1 \in I\}$. Then $I^\dagger$ is the kernel of the homomorphism $G \to K[G]/I$ and $I^\dagger \triangleleft G$. We can now state that $H = \mathsf{N}_G(T^\dagger)$ and, more properly, H is the normalizer of the unique isolated orbital subgroup of finite index above $T^\dagger$.

Note that any standard prime is induced from a prime ideal of $K[\Delta]$. This leads to our last definition. Let $N \subseteq G$ and set $H = \mathsf{N}_G(N)$. Then we let $\nabla_G(N)$ be the subgroup of G with $H \supseteq \nabla_G(N) \supseteq N$ and $\nabla_G(N)/N = \Delta(H/N)$. We can now state the following three results of [6]. The first of course is based on the work of [9].

THEOREM 10 (EXISTENCE) [6, 9]. *Let G be polycyclic-by-finite and let P be a prime ideal of $K[G]$. Then there exists an isolated orbital subgroup N of G and a prime L of $K[\nabla_G(N)]$ with $|N : L^\dagger| < \infty$ and $P = L^G$.*

Note that $\nabla_G(N)/N$ is torsion-free abelian so $\nabla_G(N)/L^\dagger$ is finite-by-abelian (and center-by-finite). Thus L is essentially a prime of a commutative group algebra. We call N above a vertex of P and L a source.

THEOREM 11 (UNIQUENESS) [6]. *Let G be polycyclic-by-finite and let P be a prime of $K[G]$. Then the vertices of P are unique up to conjugation in G. Furthermore, if N is a vertex, then the sources for this N are unique up to conjugation by $\mathsf{N}_G(N)$.*

THEOREM 12 (CONVERSE) [6]. *Let N be an isolated orbital subgroup of the polycyclic-by-finite group G. If L is a prime ideal of $K[\nabla_G(N)]$ with $|N : L^\dagger| < \infty$, then L^G is prime.*

It is a consequence of the last result that every isolated orbital subgroup is a vertex. Thus if all primes are image standard, then all isolated orbitals are normal and G is orbitally sound. The above has been extended to

COROLLARY 13 [1]. *Let N be an isolated orbital subgroup of the polycyclic-by-finite group G. If L is a prime ideal of $K[N]$ having only finitely many conjugates under $N_G(N)$, then L^G is prime.*

In closing we briefly mention the beautiful results on finite normalizing extensions. These deserve a much more detailed discussion than we can offer here. Let $S \supseteq R$ be rings with the same 1. Then S is a finite normalizing extension of R if there exist $s_1, s_2, \ldots, s_n \in S$ with $S = \sum_1^n Rs_i$ and $s_i R = Rs_i$ for all i. For example, we could have $S = R * G$ with G a finite group or S could be a finite centralizing extension where each s_i centralizes R. The following result was contributed to by a number of people. But by far the most difficult part is the incomparability due to Heinicke and Robson. So we credit all of it to their paper.

THEOREM 14 [3]. *Let $S = \sum_1^n Rs_i$ be a finite normalizing extension of R.*

(i) (*Cutting Down*) *If P is a prime ideal of S, then $P \cap R = Q_1 \cap Q_2 \cap \cdots \cap Q_t$ is an intersection of $t \leq n$ minimal covering primes of R. Furthermore, all R/Q_i are isomorphic.*

(ii) (*Lying Over*) *Let Q be a prime ideal of R. Then there exist primes $P_1, P_2, \ldots, P_s$ of S with $1 \leq s \leq n$ such that Q is minimal over each $P_i \cap R$.*

(iii) (*Incomparability*) *Let P be a prime ideal of S and let $I \triangleleft S$ with $I > P$. Then $I \cap R > P \cap R$.*

A continuation of [3] considers nilpotent ideals of S. Furthermore, the authors study intermediate extensions, that is, rings T contained between R and S. They obtain a strong relationship between the primes of R and those of T. However, they are less successful relating those of S and T. We remark that not every finite extension of interest is an intermediate extension. For example, let R be a local commutative algebra of characteristic $p > 0$ with $J = JR \neq 0$ and suppose δ is a derivation of R with $\delta^p = 0$ and $\delta J \not\subseteq J$. Then $R[x; \delta | x^p = 0]$ is not an intermediate extension of R.

References

1. K. A. Brown, *The structure of modules over polycyclic groups*, Math. Proc. Cambridge Philos. Soc. **89** (1981), 257–283.
2. J. W. Fisher and S. Montgomery, *Semi-prime skew group rings*, J. Algebra **52** (1978), 241–247.
3. A. G. Heinicke and J. C. Robson, *Normalizing extensions: prime ideals and incomparability*, J. Algebra **72** (1981), 237–268.
4. M. Lorenz and D. S. Passman, *Centers and prime ideals in group algebras of polycyclic-by-finite groups*, J. Algebra **57** (1979), 355–386.
5. ———, *Prime ideals in crossed products of finite groups*, Israel J. Math. **33** (1979), 89–132; Addendum, **35** (1980), 311–322.

6. ____, *Prime ideals in group algebras of polycyclic-by-finite groups*, Proc. London Math. Soc. (3) **43** (1981), 520–543.
7. D. S. Passman, *It's essentially Maschke's theorem*, Rocky Mountain J. Math. **13** (1983), 37–54.
8. ____, *Semiprime crossed products*, Houston J. Math. **11** (1985), 257–267.
9. J. E. Roseblade, *Prime ideals in group rings of polycyclic groups*, Proc. London Math. Soc. (3) **36** (1978), 385–447.

6. Crossed Products of Infinite Groups

Let us move on to infinite groups and consider a number of natural questions. To start with, when is $R * G$ prime or semiprime? In the case of group algebras, the main technique for this is the Δ-method and the answer depends upon the finite normal subgroups of G. To understand what is happening, we begin by assuming that R is prime.

PROPOSITION 1 [6]. *Let R be prime and let $0 \neq I \triangleleft R * G$. Then $I \cap (R * G_{\mathrm{inn}}) \neq 0$.*

PROOF. Let $0 \neq \gamma \in I$ with minimal support size and say $\operatorname{Supp} \gamma = \{1 = x_0, x_1, \ldots, x_n\}$. For each i, define
$$A_i = \left\{ r \in R \mid \sum_j r_j \bar{x}_j \in I \text{ and } r = r_i \right\}.$$

Then $0 \neq A_i \triangleleft R$. Set $A = A_0$ and note that if $\sum_i a_i \bar{x}_i \in I$, then a_0 uniquely determines the other a_i. Thus there exist well defined maps $f_i \colon A \to A_i$ given by $a_i = a_0 f_i$. These are clearly left R-module homomorphisms and they have inverses $g_i \colon A_i \to A$ defined similarly. Thus each f_i determines a unit $q_i \in Q_l(R) = S$. In addition, for all $r \in R$ we have $(a_0 r) f_i = (a_0 f_i) r^{\bar{x}_i^{-1}}$ so $q_i^{-1} r q_i = r^{\bar{x}_i^{-1}}$ and hence $x_i \in G_{\mathrm{inn}}$ for each i.

This can be used to get a weakened version of the following theorem rather quickly. The idea is to first drop down to $R * G_{\mathrm{inn}} \subseteq S * G_{\mathrm{inn}}$ and note that $S * G_{\mathrm{inn}} = S \otimes C^t[G_{\mathrm{inn}}]$. Then the known results on the primeness or semiprimeness of twisted group algebras can apply. However, the sharper result below requires a somewhat more complicated approach.

Let G permute the ideals of R. Then R is said to be a G-prime ring if for all nonzero G-invariant ideals A, B of R we have $AB \neq 0$. Similarly, R is G-semiprime if for all nonzero G-invariant ideals A of R we have $A^2 \neq 0$. Note that if $R * G$ is given and $N \triangleleft G$, then G does permute the ideals of $R * N$ by conjugation.

THEOREM 2 [7]. *Let $R * G$ be given with R a prime ring. Then $R * G$ is prime (respectively, semiprime) if and only if for all finite normal subgroups N of G with $N \subseteq G_{\mathrm{inn}}$ the ring $R * N$ is G-prime (respectively, G-semiprime).*

Note that for R prime and N finite, $R*N$ has a unique largest nilpotent ideal. Thus G-semiprime and semiprime are equivalent here. In any case, we see that the property again depends on the finite normal subgroups.

Next we move on to semiprime coefficient rings and here we discover the following interesting (that is nasty) example. Let $H \subseteq G$ and assume that the group algebra $K[H]$ has ideals $0 \neq A, B$ with $AB = 0$. Now G permutes the set Ω of right cosets of H and if $\alpha \in \Omega$ is the coset H itself, then the stabilizer G_α is also equal to H. Define the commutative semiprime ring $R = \prod_{\omega \in \Omega} K_\omega$ to be the complete direct product of copies of K indexed by the elements of Ω. Clearly G acts on R by permuting the factors and we can form the skew group ring RG.

Let e_ω denote the idempotent in R having a one in the ω-factor and zeros elsewhere. Since $e_\omega{}^g = e_{\omega g}$ it follows that, in RG, we have $e_\alpha g e_\alpha = e_\alpha g$ or 0 according to whether $g \in H$ or not. Now $K_\alpha = e_\alpha K$ so $RG \supseteq e_\alpha KH = e_\alpha K[H]$ and we can define $\tilde{A} = (RG)e_\alpha A(RG)$ and $\tilde{B} = (RG)e_\alpha B(RG)$. The above then yields $e_\alpha A g e_\alpha B = 0$ for all $g \in G$ since $A, B \triangleleft K[H]$ with $AB = 0$. Thus $\tilde{A}\tilde{B} = 0$.

In other words, RG can fail to be prime or semiprime even if G has no nontrivial finite normal subgroups. Let us look further. Set $I = K_\alpha \triangleleft R$. Then G_I, the stabilizer of I in G, is equal to H and for all $g \in G \backslash H$ we have $I^g I \subseteq I^g \cap I = 0$. Moreover, $K[H]$ is not prime (or semiprime) so H must have a finite normal subgroup N with appropriate properties. As we will see, this actually describes what happens in the case of a general crossed product $R*G$.

Here is at least an idea of the beginning of the argument. Suppose I, J are nonzero ideals of $R*G$ with $IJ = 0$. Let T be a finite subset of G with $I \cap (R*T) \neq 0$ and $J \cap (R*T) \neq 0$. We can assume that $1 \in T$ and that $A = \operatorname{tr}(I \cap R*T) \neq 0$ and $B = \operatorname{tr}(J \cap R*T) \neq 0$. Here, of course, tr indicates the identity coefficient. Clearly $A, B \triangleleft R$. Now take $g \in G$. If $A^{\bar{g}} B \neq 0$, then we can choose $\alpha = \sum a_x x \in I \cap R*T$ and $\beta = \sum b_x x \in J \cap R*T$ with $a_1{}^{\bar{g}} b_1 \neq 0$. But $\alpha^{\bar{g}} \beta = 0$, so by computing the identity coefficient we see that there exist $x, y \in T \backslash 1$ with $x^g y = 1$. Hence g must belong to an appropriate coset of $\mathbb{C}_G(x)$. Looking backwards, this says that if g is not contained in this finite union of cosets, then $A^{\bar{g}} B = 0$. Note that the latter cannot happen if R is prime and that is why the prime case is easier. In general, to deal with equations of this nature we have the following technical lemma.

LEMMA 3. *Let G act on R, let H be a subgroup of G of finite index and let $0 \neq J \triangleleft R$. Assume that $J^g J = 0$, $\forall g \in H \backslash \bigcup_1^n g_k H_k$, where the latter is a fixed finite union of cosets of the subgroups $H_k \subseteq H$. Then there exists a subgroup L of G and a nonzero product*

$$0 \neq I = J^{y_1} J^{y_2} \cdots J^{y_r} \subseteq J$$

of G-conjugates of J with $I^g I = 0$, $\forall g \in G \backslash L$. Moreover, $|L : L \cap H_k| < \infty$ for some k.

To appreciate some of the content of this lemma, suppose we are given $|G : H| < \infty$ and $H = \bigcup_1^n g_k H_k$. Let G act trivially on any field K and take

$J = K$. Then $J^g J = 0$ for all $g \in H \backslash \bigcup_1^n g_k H_k$, since the latter set difference is empty, and thus the above applies. But then clearly we must have $G \backslash L = \emptyset$ so $L = G$ and $|G : H_k| < \infty$ for some k. In other words, this lemma generalizes the group theoretic property we considered earlier.

The main result is

THEOREM 4 [9]. *Given the crossed product $R*G$, there exist ideals $0 \neq A, B$ of $R*G$ with $AB = 0$ if and only if there exist*
(i) *$N \triangleleft H \subseteq G$ with N finite,*
(ii) *an H-invariant ideal $0 \neq I$ of R with $I^{\bar g} I = 0$ for all $g \in G \backslash H$,*
(iii) *H-invariant ideals $0 \neq \tilde A, \tilde B$ of $R*N$ with $\tilde A, \tilde B \subseteq I*N$ and $\tilde A \tilde B = 0$.*
Moreover, $A = B$ if and only if $\tilde A = \tilde B$.

Note that if $R*G$ is prime or semiprime, then R is G-prime or G-semiprime respectively. Thus there is no loss in assuming that R is G-semiprime. In that case, the above result can take on a slightly different form. We say that $I \triangleleft R$ is a T.I. ideal (trivial intersection ideal) if for all $g \in G$, either $I^{\bar g} = I$ or $I^{\bar g} \cap I = 0$. We state two corollaries which use these ideals and finesse the finite group problem.

COROLLARY 5 [9]. *Let $R*G$ be given with R a G-prime ring. Assume that for all T.I. ideals $0 \neq I \triangleleft R$ we have $\Delta^+(G_I) = 1$. Then $R*G$ is prime.*

In particular, if G is torsion-free and R is G-prime, then $R*G$ is prime.

COROLLARY 6 [9]. *Let $R*G$ be given with R a G-semiprime ring. Assume that for all T.I. ideals $0 \neq I \triangleleft R$ we have $\Delta^p(G_I) = 1$ if I has p-torsion. Then $R*G$ is semiprime.*

Thus if R has characteristic 0 (that is has no torsion) and if R is G-semiprime, then $R*G$ is semiprime.

We would of course like to go further and actually study the prime ideals P of $R*G$ and again we can assume that $P \cap R = 0$. At the very least, some version of incomparability is required. Note that in the ordinary polynomial ring $K[x_1, \ldots, x_n]$, if $0 = P_0 < P_1 < \cdots < P_{n+1}$ is a chain of prime ideals, then $P_{n+1} = K[x_1, \ldots, x_n]$ so $P_{n+1} \cap K = K > P_0 \cap K = 0$. The best one can hope for is an analogous result for crossed products with G polycyclic-by-finite. In this case $R*G$ can be viewed as a generalized polynomial ring where the number of variables corresponds to the Hirsch number $h(G)$. Recall that $h(G)$ is the number of infinite cyclic factors in a subnormal series for G and that any finitely generated nilpotent group is polycyclic. We know at least

THEOREM 7 [2]. *Let $R*G$ be given with G a finitely generated nilpotent group of Hirsch number n. If $P_0 < P_1 < \cdots < P_k$ is a chain of primes in $R*G$ with $k > 2^n$, then $P_k \cap R > P_0 \cap R$.*

This of course yields information on the prime length of $R*G$ in terms of $h(G)$ and the G-prime length of R. The general case of polycyclic-by-finite groups is still open.

We now change topics slightly and discuss group-graded rings. The notation here may be slightly confusing. Let G be a multiplicative group. The ring R is G-graded if $R = \oplus \sum_{x \in G} R(x)$ is a direct sum of additive subgroups $R(x)$ indexed by the elements $x \in G$, and if $R(x)R(y) \subseteq R(xy)$. Clearly $R(1)$ is a subring, the base ring, and we have $1 \in R(1)$. We say R is strongly graded if $R(x)R(y) = R(xy)$ for all $x, y \in G$.

If X is any subset of G, we let $R(X) = \sum_{x \in X} R(x)$. Thus $R(G) = R$, and if H is a subgroup of G, then $R(H)$ is the naturally contained H-graded subring. Here are some examples.

1. Let S be the crossed product $S = R * G$. Then $S(x) = R\bar{x}$, $S(1) = R$, so R is the base ring and $S(H) = R * H$.

2. Let $I \triangleleft R$ and assume that $\bigcap_0^\infty I^n = 0$. Then there is a natural Z-grading on the associated graded ring $S = \oplus \sum_0^\infty I^n/I^{n+1}$.

3. Let G be a finite group and let $M_G(R)$ denote the ring of $|G| \times |G|$ matrices over R. We index the rows and columns by the elements of G and assign a grade of $x^{-1}y$ to the entries in the (x,y)-position. In this way, $M_G(R)$ becomes G-graded with base ring the set of diagonal matrices. Note that if G is infinite we can still introduce a grading on the row finite matrices. However, this does not make $M_G(R)$ a G-graded ring since certain infinite sums arise.

4. Let L be a Lie algebra over K and let $U(L)$ be its universal enveloping algebra. If $\lambda\colon L \to K$ is a linear functional, let

$$S_\lambda = \{u \in U(L) | [l, u] = \lambda(l)u \text{ for all } l \in L\}.$$

These are the semi-invariants corresponding to λ. Now set $S = \sum_\lambda S_\lambda$. Then one knows that the sum is direct and that $S_\lambda S_\mu \subseteq S_{\lambda+\mu}$. Thus the semicenter S of $U(L)$ is graded by the additive group $\text{Hom}_K(L, K)$.

Group graded rings were introduced by Dade [4] as a formal way to deal with group representation problems. It was clear to him (and most group theorists) that standard module arguments would apply here. It is interesting to speculate as to whether he knew about the version of Maschke's theorem which holds. We will discuss this later.

Now we would certainly like to extend the crossed product results mentioned earlier to group-graded rings. However, it quickly becomes apparent that the old techniques do not carry over. Fortunately they are not necessary since there exists a duality machine begun in [3] and extended in [10] which translates many of the crossed product results directly to this new context. We use the simpler approach of [10] and for convenience assume that G is finite.

Let R be G-graded with G finite. Form $M_G(R)$ and for every subgroup H of G define $R\{H\} \subseteq M_G(R)$ by

$$R\{H\} = \{\alpha \in M_G(R) | \alpha_{xy} \in R(x^{-1}Hy)\}.$$

Here α_{xy} denotes the (x,y)-entry of the matrix α. Then $R\{H\}$ is a subring of the matrix ring and $R\{G\} = M_G(R)$. In addition, G embeds in $M_G(R)$ via

$g \to \bar{g} = (\delta_{xg,y})$, the permutation matrix with a 1 in the (x, xg)-entries and zeros elsewhere.

LEMMA 8. *Let H be a subgroup of the finite group G. Then*
$$R\{H\} = \oplus \sum_{h \in H} R\{1\}\bar{h} = R\{1\} * H,$$
a skew group ring.

The duality result of [3] follows from $R\{1\} * G = R\{G\} = M_G(R)$ once $R\{1\}$ is suitably identified as a smash product. That result per se has few applications. The applications really follow from the specific realization of the isomorphism. Here is a sample application. We say that the grading on R is nondegenerate if for all $x \in G$ and $0 \neq r \in R(x)$ we have $rR(x^{-1}) \neq 0$ and $R(x^{-1})r \neq 0$. If R is semiprime and G is finite, then a result of Cohen and Rowen asserts that R is nondegenerate. Conversely

THEOREM 9 [3]. *Let R be a nondegenerate G-graded ring with G finite. If the base ring $R(1)$ is semiprime with no $|G|$-torsion, then R is semiprime.*

This is of course a natural extension of the Fisher-Montgomery theorem. It is proved by studying the embedding of R into $R\{1\}$. On the other hand, [10] finesses this aspect and is able to study all subrings $R\{H\}$ directly as follows.

Write $G = \{x_1, x_2, \ldots, x_n\}$ and for any subgroup H of G define the diagonal subsets $D(H)$, $D^{-1}(H)$ of $M_G(R)$ by
$$D(H) = \text{diag}(R(Hx_1), R(Hx_2), \ldots, R(Hx_n)),$$
$$D^{-1}(H) = \text{diag}(R(x_1^{-1}H), R(x_2^{-1}H), \ldots, R(x_n^{-1}H)).$$

LEMMA 10. *If H is a subgroup of the finite group G, then*
$$D^{-1}(H) \cdot M_G(R(H)) \cdot D(H) \subseteq R\{H\},$$
$$D(H) \cdot R\{H\} \cdot D^{-1}(H) \subseteq M_G(R(H)).$$

In other words, $R\{H\}$ and $M_G(R(H))$ are *generalized conjugates* by diagonal matrices. This suffices for most applications. For example, we say that R is component regular if $\underline{r}(R(x)) = 0$ and $l(R(x)) = 0$ for all $x \in G$. More generally, an R-module V is component regular if $vR(x) = 0$ implies $v = 0$. Here is a nice application.

THEOREM 11 [10]. *Let R be a G-graded ring with G finite.*
(i) *If R is semiprime and H is a subgroup of G, then $R(H)$ is semiprime.*
(ii) *Assume that $R(1)$ is semiprime and that $R(P)$ is semiprime for all elementary abelian p-subgroups $P \subseteq G$ for which R has p-torsion. If R is component regular, then it is semiprime.*

PROOF (SKETCH). (i) Since R is semiprime so is $M_G(R) = R\{1\} * G$. The crossed product result now implies that $R\{1\} * H = R\{H\}$ is semiprime. We conclude from the generalized conjugation that $M_G(R(H))$ is semiprime and hence so is $R(H)$.

(ii) Note that the torsion of $M_G(R)$ is identical to that of R. Now $R(P)$ semiprime implies the same for $M_G(R(P))$. Next the generalized conjugation along with component regularity shows that $R\{P\} = R\{1\} * P$ is semiprime. Finally, the crossed product result implies that $R\{1\} * G = M_G(R)$ is semiprime and hence so is R.

This method also yields graded ring analogs of the results discussed earlier on necessary and sufficient conditions for $R * G$ to be prime or semiprime with G infinite. Again some minimal hypothesis on the grading is required to preserve the group structure. One can also obtain module information this way. For example, here is the essential version of Maschke's theorem.

THEOREM 12 [8, 10]. *Let R be a G-graded ring with G finite and let $W \subseteq V$ be component regular R-modules with no $|G|$-torsion. Then $W \operatorname{ess}_R V$ if and only if $W \operatorname{ess}_{R(1)} V$.*

PROOF. One direction is trivial, so assume $W \operatorname{ess}_R V$. Let $\overline{V} = \operatorname{Row}_G(V)$ be the set of row vectors of size $|G|$ with entries in V. Then $\overline{V} \supseteq \overline{W}$ and these are both right $M_G(R)$-modules. Furthermore, $\overline{W} \operatorname{ess}_{M_G(R)} \overline{V}$. Since $M_G(R) = R\{1\} * G$, the crossed product result implies that $\overline{W} \operatorname{ess}_{R\{1\}} \overline{V}$. Finally, take $v \in V \backslash 0$ and let $\overline{v} = [v, 0, \ldots, 0] \in \overline{V}$. Then $\overline{v}R\{1\} \cap \overline{W} \neq 0$ so the definition of $R\{1\}$ implies that there exists $r \in R(x)$ for some x with $0 \neq vr \in W$. Since W is component regular, we have $vrR(x^{-1}) \neq 0$ so $vR(1) \cap W \neq 0$.

A related result of interest is the following lemma of Grzeszczuk.

LEMMA 13. *Let R be a G-graded ring with G finite and let $0 \neq V$ be an R-module. Suppose W is an $R(1)$-submodule of V with $W \operatorname{ess}_{R(1)} V$. Then W contains a nonzero R-submodule of V.*

PROOF. Since $V \neq 0$ we have $W \neq 0$. Note that $w \in W \backslash 0$ implies that $wR(1) \subseteq W$. Now choose X a subset of G of maximal size such that there exists $w \in W \backslash 0$ with $wR(X) \subseteq W$. We claim that $X = G$. If not, let $g \in G \backslash X$. If $wR(g) = 0$, then $wR(X \cup \{g\}) \subseteq W$, a contradiction. Thus $wR(g) \neq 0$.

Now $wR(g)$ is a nonzero $R(1)$-submodule of V and $W \operatorname{ess}_{R(1)} V$. Thus there exists $w' \in wR(g) \cap W$ with $w' \neq 0$. Observe that
$$w'R(g^{-1}X) \subseteq wR(g)R(g^{-1}X) \subseteq wR(X) \subseteq W$$
and $w' \in W$ so $w'R(g^{-1}X \cup \{1\}) \subseteq W$, again a contradiction. Thus $X = G$ and $wR(G)$ is a nonzero R-submodule contained in W.

THEOREM 14 [5]. *Let R be a G-graded ring with G finite and let V be a completely reducible R-module. Then the restriction $V_{R(1)}$ is a completely reducible $R(1)$-module.*

PROOF. We can assume that V is irreducible. The lemma then implies that $V_{R(1)}$ has no proper essential submodules and hence it is completely reducible.

Returning to the duality, we mention another module application, this time for infinite groups.

THEOREM 15 [**11**]. *Let R be a G-graded ring with G polycyclic-by-finite and let M be a graded-Noetherian R-module. Then M is a Noetherian R-module and*

$$\text{Kdim}_R M \leq \text{gr}\,\text{Kdim}_R M + h(G).$$

This extends a result of [**8**] on nilpotent groups and generalizes a result of Bell on strongly graded rings.

We remark that the duality considered here is part of a more general duality for Hopf algebras. Unfortunately, the infinite-dimensional case is too technical to state here, but the finite-dimensional case at least gives the flavor. Here # denotes the Hopf algebra smash product.

THEOREM 16 [**1**]. *Let H be a Hopf algebra of dimension $n < \infty$ over a field K and let H^* be its dual. If A is an H-module algebra then*

$$(A\#H)\#H^* \simeq A \otimes (H\#H^*) \simeq A \otimes M_n(K) \simeq M_n(A).$$

In the infinite-dimensional case, H^* is too large so [**1**] uses a certain Hopf subalgebra U of the *finite dual* H^0. We remark that in the case of G-graded rings for G infinite, this duality does not agree with that of [**10**]. The result of [**1**] is more functorially defined but only exists for residually finite f.c. groups.

References

1. R. J. Blattner and S. Montgomery, *A duality theorem for Hopf module algebras*, J. Algebra **95** (1985), 153–172.
2. W. Chin, *Prime ideals in differential operator rings and crossed products*, J. Algebra.
3. M. Cohen and S. Montgomery, *Group-graded rings, smash products and group actions*, Trans. Amer. Math. Soc. **282** (1984), 237–258.
4. E. C. Dade, *Group graded rings and modules*, Math. Z. **174** (1980), 241–262.
5. P. Grzeszczuk, *On G-systems and G-graded rings*, Proc. Amer. Math. Soc. **95** (1985), 348–352.
6. S. Montgomery, *Outer automorphisms of semi-prime rings*, J. London Math. Soc. (2) **18** (1978), 209–220.
7. S. Montgomery and D. S. Passman, *Crossed products over prime rings*, Israel J. Math. **31** (1978), 224–256.
8. C. Nastasescu, *Group rings of graded rings, applications*, J. Pure Appl. Algebra **33** (1984), 313–335.
9. D. S. Passman, *Infinite crossed products and group-graded rings*, Trans. Amer. Math. Soc. **284** (1984), 707–727.
10. D. Quinn, *Group-graded rings and duality*, Trans. Amer. Math. Soc. **292** (1985), 155–167.
11. ____, Thesis, University of Wisconsin, 1985.

7. Computing the Symmetric Ring of Quotients

Let R be a prime ring. We recall that $Q_l(R)$, the left Martindale ring of quotients of R, is defined as the set of equivalence classes of left module homomorphisms $f\colon {}_RA \to {}_RR$ where A runs over the nonzero two-sided ideals of R. Therefore, the following abstract characterization comes as no surprise.

PROPOSITION 1. *Let R be prime. Then $Q_l = Q_l(R)$ has the following four properties.*

(i) $Q_l \supseteq R$ *with the same* 1.
(ii) *If $q \in Q_l$, then there exists $0 \neq A \triangleleft R$ with $Aq \subseteq R$.*
(iii) *If $q \in Q_l$ and $0 \neq A \triangleleft R$, then $Aq = 0$ implies $q = 0$.*
(iv) *Let $0 \neq A \triangleleft R$ and let $f\colon {}_RA \to {}_RR$ be an R-module homomorphism. Then there exists $q \in Q_l$ with $aq = af$ for all $a \in A$.*

Furthermore, Q_l is uniquely characterized by these properties.

Of course, the right ring of quotients $Q_r(R)$ also exists and has an analogous characterization. Here are some examples of Q_l.

1. Let R be a simple ring. Then $Q_l(R) = R$. Indeed (i), (ii) and (iii) are obvious and (iv) follows since the only left module maps $f\colon {}_RR \to {}_RR$ are right multiplication by elements of R.

2. Let R be a commutative domain with field of fractions F. Then $Q_l(R) = F$. Again (i), (ii) and (iii) are trivially satisfied by F. For (iv), suppose $f\colon {}_RA \to {}_RR$ is given. If $a, b \in A\backslash 0$, then
$$a(bf) = (ab)f = (ba)f = b(af)$$
so $a^{-1}(af) = b^{-1}(bf)$ and we see that $a^{-1}(af)$ is the same element $q \in F$ for all $a \in A\backslash 0$.

3. Let $R = K\langle x, y \rangle$ be the free algebra on two variables and let I be the ideal generated by x and y. Then $I = Rx + Ry$ is the free R-module on x and y and we can define $f\colon {}_RI \to {}_RR$ by $xf = 0$ and $yf = 1$. Thus, by (iv), there exists $q \in Q_l(R)$ with $xq = 0$ and $yq = 1$. In particular, $Q_l(R)$ has zero divisors even though R itself is a domain.

It is because of example (2) that Q_l is a ring of quotients and it is because of (3) that Q_l is really too large.

Martindale [3] introduced Q_l to study generalized polynomial identities. For the most part, he restricted his attention to the smaller ring RC where $C = Z(Q_l)$

is the extended centroid of R. RC is called the central closure of R. It is prime and centrally closed. Martindale used this extension ring to handle certain types of linear identities. In particular he proved (i) below and (ii) is an easy consequence.

PROPOSITION 2 [3]. *Let R be a prime ring and C its extended centroid.*

(i) *Let $a_i, b_i \in R\backslash 0$ and suppose $\sum_1^n a_i x b_i = 0$ for all $x \in R$. Then $a_1, a_2, \ldots, a_n$ are linearly dependent over C.*

(ii) *Assume that R is centrally closed and let E be an algebra over C. If P is a prime ideal of $R \otimes_C E$ with $P \cap R = 0$, then $P = R \otimes Q$ with Q a prime ideal of E.*

The following computations concern group rings. Here Q_{cl} denotes the classical ring of quotients.

THEOREM 3 [1, 8]. *Let $K[G]$ be a prime group ring with center Z.*
(i) *If $K[G]$ is an Ore ring, then $Z(Q_{cl}(K[G])) = Q_{cl}(Z)$.*
(ii) $Q_{cl}(Z) = C$, *the extended centroid of $K[G]$.*

Kharchenko observed that Q_l can be used to develop a Galois theory of prime rings. He actually worked with semiprime rings (there is an appropriate definition of Q_l in this case) but the subject becomes much more unpleasant. Recall that $\sigma \in \operatorname{Aut} R$ is X-inner if its unique extension to Q_l is inner. A group G of automorphisms is X-outer if only the identity is X-inner. He showed

LEMMA 4. *Let R be a prime ring with $a_i, b_i \in R$ and $\sigma_i \in \operatorname{Aut} R$ for $i = 0, 1, \ldots, n$. Assume that $a_0, b_0 \neq 0$ and $\sigma_0 = 1$. If $\sum_0^n a_i x^{\sigma_i} b_i = 0$ for all $x \in R$, then, for some $i \neq 0$, σ_i is X-inner.*

COROLLARY 5 [2]. *If G is a group of X-outer automorphisms of R, then the elements of G are linearly independent automorphisms.*

The independence of automorphisms is of course a key ingredient in the ordinary Galois theory of fields. This led Montgomery to define the normal closure RN of R to be the subring of Q_l generated by R and

$$N = \{\text{units } q \in Q_l | q^{-1} R q = R\}.$$

That is, N is the set of units which define X-inner automorphisms. It follows that RN is an extension ring of R which is still prime, but it is not normally closed in general (example later). Since RN is large enough for the study of crossed products and Galois theory, numerous special cases have been computed. We start with

THEOREM 6 [5]. *Let A be a filtered algebra, so that $A = \bigcup_{n=0}^{\infty} A_n$ with $A_n A_m \subseteq A_{n+m}$, and assume that the associated graded algebra $\overline{A}$ is a commutative domain. If σ is an X-inner automorphism of $\overline{A}$, then σ preserves the filtration and acts trivially on $\overline{A}$.*

PROOF (SKETCH). If $a \in A_n \backslash A_{n-1}$, let $\overline{a} \in A_n/A_{n-1} \subseteq \overline{A}$ be its *leading term*. Since $\overline{A}$ is a domain it follows that $\overline{ab} = \overline{a}\overline{b}$. Now let σ be an X-inner automorphism. Then there exist $a, b, c, d \in A\backslash 0$ with $arb = cr^\sigma d$ for all $r \in A$. We therefore have $\overline{a}\,\overline{r}\overline{b} = \overline{c}\overline{r^\sigma}\overline{d}$ for all $r \neq 0$. In particular, setting $r = 1$ yields $\overline{a}\overline{b} = \overline{c}\overline{d}$ and thus since $\overline{A}$ is a commutative domain we see that $\overline{r} = \overline{r^\sigma}$. Thus σ preserves the filtration and acts trivially on $\overline{A}$.

This of course applies to $U = U(L)$, the universal enveloping algebra of a Lie algebra L over K. Indeed we know that U is filtered and that $\overline{U}$ is a commutative polynomial ring. Also, if L is finite-dimensional, then U is an Ore domain.

Let q be a unit of $Q_l(U)$ which gives rise to an X-inner automorphism σ. Then the above implies that $q^{-1}lq = l + \lambda(l)$ for all $l \in L$ where $\lambda: L \to K$. Thus $\lambda \in \mathrm{Hom}(L, K)$ and $[l, q] = lq - ql = \lambda(l)q$ so q is a semi-invariant for L with $q \in Q_l(U)_\lambda$. Conversely, suppose $0 \neq q \in Q_l(U)$ with $[l, q] = \lambda(l)q$ for all $l \in L$. It follows that $U(L)q = qU(L)$ and hence that q is a unit of $Q_l(U)$. Multiplying by q^{-1} then yields $q^{-1}lq = l + \lambda(l)$ and again we have an X-inner automorphism. We conclude that

COROLLARY 7 [5]. *Let L be a Lie algebra over K and let $U = U(L)$ be its enveloping algebra.*

(i) *The group of X-inner automorphisms of $U(L)$ is isomorphic to the additive subgroup of $\mathrm{Hom}(L, K)$ of those λ with $Q_l(U)_\lambda \neq 0$.*

(ii) *The semi-invariants for L in U are precisely the normal elements of U. Hence, the semicenter is a characteristic subring of U.*

Filtered rings occur naturally in the study of coproducts. Let R_1 and R_2 be rings containing a common division ring D. Then $R = R_1 \amalg R_2$, the coproduct over D, is filtered by $F^0 = D$ and $F^n = (R_1 + R_2)^n = \sum R_{i_1} R_{i_2} \cdots R_{i_n}$. The X-inner automorphisms of such rings have been studied in a series of papers by Lichtman, Martindale and Montgomery (in various combinations). The best result now is

THEOREM 8 [4]. *Assume that each $R_i > D$, at least one of the dimensions over D is larger than 2, and one-sided inverses in R_i are two-sided. Then every X-inner automorphism of $R = R_1 \amalg R_2$ is inner unless one of the following occurs.*

(i) *Each R_i is primary, that is, $R_i = D + T_i$ with $T_i^2 = 0$.*

(ii) *One R_i is primary and the other is 2-dimensional.*

(iii) *$\mathrm{char}\, D = 2$, one R_i is not a domain, and one is quadratic.*

In the course of the proof one shows that X-inner automorphisms σ are strongly bounded; that is, there exists an integer $k \geq 0$ with $\deg r^\sigma \leq \deg r + k$ for all $r \in R$.

As we observed above, if q induces an X-inner automorphism of $U(L)$, then q normalizes $L + K$. As an analog in crossed products we consider those X-inner automorphisms of $R * G$ which normalize the group of trivial units. The following result gives the idea without all the tedious details. Recall that the condition $G_{\mathrm{inn}} \cap \Delta^+ = 1$ implies that $R * G$ is prime.

THEOREM 9 [6]. *Let $R * G$ be a crossed product with R prime and $G_{\text{inn}} \cap \Delta^+ = 1$. Suppose q is a unit of $Q_l(R * G)$ and $\sigma \in \text{Aut } G$ with*

$$q^{-1} R \overline{x} q = R \overline{x^\sigma} \qquad \forall x \in G.$$

Then $\sigma = \sigma_1 \sigma_2$ where σ_1 centralizes a subgroup of G of finite index and σ_2 is an inner automorphism of G. Furthermore, there exists a fairly tight description of the element q.

COROLLARY 10 [6]. *Let $R * G$ be given with R prime and $G_{\text{inn}} \cap \Delta^+ = 1$ and let $S * G$ be the unique extension of $R * G$ with $S = Q_l(R)$. Suppose X is the group of X-inner automorphisms of $R * G$ which normalize both R and the group of trivial units of $R * G$. If X_0 is the subgroup of X consisting of those automorphisms induced by trivial units of $S * G$, then X/X_0 is a periodic abelian group.*

We remark that we frequently see theorems which assert that X-Inn $R/\text{Inn } R$ is a periodic abelian group. This is certainly a special situation since, for domains R, X-Inn $R/\text{Inn } R$ can be any group.

Let us return again to enveloping algebras and use them as the coefficient ring of a skew group ring $H = U(L)G$. Then H is actually a co-commutative Hopf algebra. Conversely, a result of Kostant asserts that every co-commutative Hopf algebra over an algebraically closed field of characteristic 0 is of this form. Moreover, the Hopf algebra structure essentially picks out L and G.

Note that since $U(L)$ is a domain and we know its group of X-inner automorphisms, it is easy to determine when H is prime. For example, in characteristic 0 this occurs if and only if $\mathbb{C}_G(U(L)) \cap \Delta^+ = 1$.

Let $0 \neq q \in Q_l(H) = Q_l(U(L)G)$. We say that q is a semi-invariant for L and G if and only if there exists $\mu: L \to K$ with

$$[l, q] = \mu(l) q \qquad \forall l \in L$$

and there exists $\lambda: G \to K \backslash 0$ with

$$x^{-1} q x = \lambda(x) q \qquad \forall x \in G.$$

That is, q is a common eigenvector for the natural actions of L and of G on Q_l. This actually makes sense in Hopf algebra terms. Indeed q is a common eigenvector for the Hopf inner action of H on $Q_l(H)$.

Note that if q is a semi-invariant, then q is a unit of $Q_l(H)$ normalizing $U(L)$ and the group of trivial units of $U(L)G$. Thus the above applies. That along with known results on the semi-invariants of $U(L)$ then yields

COROLLARY 11 [6]. *Let $U(L)G$ be a skew group ring of G over the universal enveloping algebra $U(L)$ of the finite K-dimensional Lie algebra L. Assume that $H = U(L)G$ is prime and that char $K = 0$ and let SZ denote the linear span of all semi-invariants for L and G contained in H.*

(i) *SZ is a commutative integral domain.*

(ii) *Every semi-invariant for L and G in $Q_l(H)$ is a quotient of semi-invariants in SZ.*

As we said earlier, the normal closure RN is large enough to handle problems in crossed products and Galois theory. But it is not large enough for derivations. Let $\delta \colon R \to R$ be a derivation. Then δ extends uniquely to a derivation $\delta \colon Q_l(R) \to Q_l(R)$. As usual, we say that δ is X-inner if it becomes inner on Q_l, that is, if there exists $q \in Q_l$ with $\delta(r) = [r, q] = rq - qr$ for all $r \in R$ (or $r \in Q_l(R)$). This concept is needed, for example, in

LEMMA 12. *Let R be a prime algebra over the rationals and let $R[x; \delta]$ be a differential operator ring in the variable x. If δ is not X-inner, then for any ideal $0 \neq I \triangleleft R[x; \delta]$ we have $I \cap R \neq 0$.*

If δ is an X-inner derivation induced by $q \in Q_l(R)$, then q need not belong to the normal closure RN. However it does belong to a larger and still well behaved subring of Q_l. Rather than merely adding all such q to RN, we take a slightly different point of view.

Let R be a prime ring. We have already defined $Q_l(R)$ and its right analog $Q_r(R)$. In view of Proposition 1, we might expect a symmetric Martindale ring of quotients to be characterized as follows.

PROPOSITION 13. *Let R be prime. Then the ring $Q_s = Q_s(R)$ is uniquely determined by the following four properties.*
  (i) $Q_s(R) \supseteq R$ *with the same* 1.
  (ii) *If $q \in Q_s$, then there exist $0 \neq A, B \triangleleft R$ with $Aq, qB \subseteq R$.*
  (iii) *Let $q \in Q_s$ and let $0 \neq I \triangleleft R$. Then $qI = 0$ or $Iq = 0$ implies that $q = 0$.*
  (iv) *Let $0 \neq A, B \triangleleft R$ and suppose $f \colon {}_RA \to {}_RR$ and $g \colon B_R \to R_R$ are module homomorphisms satisfying*

$$(af)b = a(gb) \qquad \forall a \in A, b \in B.$$

Then there exists $q \in Q_s$ with $af = aq$ and $gb = qb$ for all $a \in A$, $b \in B$.

Notice that the balanced or associative condition $(af)b = a(gb)$ is necessary in (iv) since it is equivalent to the associativity statement $(aq)b = a(qb)$ in Q_s. Notice also that while Q_s is unique, at this point we do not know that it exists. One approach, of course, is to consider the set of all equivalence classes of pairs (f, g) of balanced maps. However, we avoid this by identifying Q_s as a subring of Q_l.

PROPOSITION 14. *If R is a prime ring, then $Q_s(R)$ exists. Indeed*

$$Q_s(R) = \{q \in Q_l | qB \subseteq R \text{ for some } 0 \neq B \triangleleft R\}$$

and

$$Q_s(R) = \{q \in Q_r | Aq \subseteq R \text{ for some } 0 \neq A \triangleleft R\}.$$

PROOF. Set $S = \{q \in Q_l | qB \subseteq R \text{ for some } 0 \neq B \triangleleft R\}$. Since R is prime, it follows easily that S is a subring of $Q_l(R)$ containing R. The goal is to show that

S satisfies (i)–(iv) of Proposition 13. Of course, (i) and (ii) follow by definition and for (iii) we know at least, since $q \in Q_l$, that $Iq = 0$ implies $q = 0$. In the other direction, if $qI = 0$, then choose $0 \neq A \triangleleft R$ with $Aq \subseteq R$. Then $(Aq)I = 0$ so $Aq = 0$ since R is prime and hence $q = 0$. Finally, let f, g be given as in (iv). Then we know at least that there exists $q \in Q_l$ with $af = aq$ for all $a \in A$. The balanced condition and associativity now imply that $a(gb) = (af)b = (aq)b = a(qb)$ so $A(gb - qb) = 0$ and hence $gb = qb$ for all $b \in B$. Finally, $qB = gB \subseteq R$ so $q \in S$.

This subring of Q_l was used by Kharchenko in his work on Galois theory, although he did not really stress the symmetric aspects of the ring. The symmetric formulation can presumably simplify some arguments—my later work with Montgomery, in particular.

Two remarks are now in order. First, if R is a domain, then so is $Q_s(R)$. Thus Q_s is not *too big*. Second, $Q_s(R) \supseteq RN$ and it contains all q inducing X-inner derivations. Thus it is *big enough*. Let us consider some specific computations.

1. Let $R = K\langle x, y, \ldots \rangle$ be a free algebra on at least two generators. Then $Q_s(R) = R$. More generally Kharchenko showed that if R satisfies a 2-term weak algorithm and if $R \neq D[x; \sigma]$, a skew polynomial ring over a division ring, then $Q_s(R) = R$.

2. This is the example used by Bergman to show that RN need not be normally closed. Let $R = K[t][x, y | xy = tyx]$. Then we have

$$Q_l(R) = K(t)[x^{-1}, x, y | xy = tyx],$$
$$Q_r(R) = K(t)[x, y, y^{-1} | xy = tyx],$$
$$Q_s(R) = K(t)[x, y | xy = tyx].$$

Thus $Q_s(R) = RC = RN$. Also if S is any of the three rings above, then

$$Q_l(S) = Q_r(S) = Q_s(S) = K(t)[x^{-1}, x, y, y^{-1} | xy = tyx],$$

a simple ring. This shows that none of Q_l, Q_r, Q_s, RN is a closure operator. On the other hand, these examples terminate in two steps. One wonders whether there is an example which keeps growing for infinitely many steps.

3. Let $M_\infty(K)$ be the set of all $\infty \times \infty$ matrices over K. Then this is not a ring but it does contain the rings of row finite matrices and of column finite matrices. Let I be the set of all finite matrices (that is having only finitely many nonzero entries) and let $R = K + I$. Then

$$Q_l(R) = \{\text{row finite matrices}\},$$
$$Q_r(R) = \{\text{column finite matrices}\},$$
$$Q_s(R) = \{\text{row and column finite matrices}\}.$$

One would like to say in general that $Q_s = Q_l \cap Q_r$. However this example shows that Q_l and Q_r are not subrings, in a natural manner, of a common ring.

Now let us look at group rings. First, $K[G]$ must be prime so $\Delta^+(G) = 1$. Next we have

LEMMA 15. *If $Q_s(K[G]) = K[G]$, then $\Delta(G) = 1$.*

This is proved by first showing that $Z(K[G])$ must be trivial. Thus if $K[G]$ is symmetrically closed, then G has no nonidentity finite conjugacy classes. It turns out that the countable classes are the interesting ones since

THEOREM 16 [7]. *If all nonidentity conjugacy classes of G are uncountable, then $K[G]$ is symmetrically closed.*

What about the countable classes? They each generate, of course, a countable normal subgroup of G and we have at least

PROPOSITION 17 [7]. *Let $\Delta^+(G) = 1$ and let $1 \neq N$ be a countable locally finite normal subgroup of G. Then $K[G]$ is not symmetrically closed.*

COROLLARY 18 [7]. *Let G be a locally finite group with $\Delta^+(G) = 1$. Then $K[G]$ is symmetrically closed if and only if all nonidentity conjugacy classes of G are uncountable.*

Let $0 \neq I \triangleleft K[G]$ and let $H \triangleleft G$. Then an intersection theorem asserts that under suitable hypotheses $I \cap K[H] \neq 0$. Such theorems can be used to compute $Q_s(K[G])$. We will skip the precise details here and just mention some groups to which they apply.

THEOREM 19 [7]. *(i) If G is a nonabelian free group or an algebraically closed group, then $K[G]$ is symmetrically closed.*

(ii) If G is a polycyclic-by-finite group with $\Delta^+(G) = 1$, then $Q_s(K[G]) = Z^{-1}K[G]$, the central closure of $K[G]$, where $Z = Z(K[G])$.

We remark that if R is a prime Noetherian ring, then $Q_s(R)$ need not equal the central closure RC. Thus part (ii) above does have content. Finally, it appears that the problem of finding necessary and sufficient conditions for $K[G]$ to be symmetrically closed will prove to be elusive. Aspects of it seem to be related to the zero divisor problem.

References

1. E. Formanek, *Maximal quotient rings of group rings*, Pacific J. Math. **53** (1974), 109–116.
2. V. K. Kharchenko, *Generalized identities with automorphisms*, Algebra i Logika **14** (1975), 215–237; English transl. 1976, 132–148.
3. W. S. Martindale, *Prime rings satisfying a generalized polynomial identity*, J. Algebra **12** (1969), 576–584.
4. ____, *The normal closure of the coproduct of rings over a division ring*.
5. S. Montgomery, *X-inner automorphisms of filtered algebras*, Proc. Amer. Math. Soc. **83** (1981), 263–268.
6. S. Montgomery and D. S. Passman, *X-inner automorphisms of crossed products and semi-invariants of Hopf algebras*, Israel J. Math.
7. D. S. Passman, *Computing the symmetric ring of quotients*, J. Algebra.
8. M. K. Smith, *Group algebras*, J. Algebra **18** (1971), 477–499.

8. Galois Theory and Crossed Products

Let G act on the ring R so that we have a group homomorphism $G \to \operatorname{Aut} R$. Then we can form the skew group ring RG which is associative and contains R, G and the fixed ring R^G. In other words, RG contains all the ingredients for the study of the Galois theory of rings. Thus we might hope to use crossed product results to obtain Galois theoretic information. Indeed there exist such applications, but perhaps not an overwhelming amount for the following reasons.

First, it seems necessary for us to assume G is finite. The reason is that only certain classes of infinite groups are allowed (discussed in the next section) and their description does not readily translate into crossed product terms. Second, the structure of RG is best understood when R has no $|G|$-torsion. Thus the route through crossed products usually requires that assumption. Third, certain natural Galois theory questions may not translate to natural questions about RG and vice versa.

In this section we will discuss skew group ring applications and we will always assume G to be finite here. We start with the existence of fixed points. Let G act on R and let I be a nonzero G-invariant ideal of R. The question is whether $I^G = I \cap R^G$ must necessarily be nonzero. The answer is *yes* and *no*. For *no* we have the following example of Bergman.

Let $S = K\langle x, y\rangle$ be the free algebra on x, y over the field K of characteristic $p > 2$ and let $R = M_2(S)$ be the ring of 2×2 matrices over S. Now let G be the group of units of R generated by the matrices

$$\begin{bmatrix} -1 & 0 \\ 0 & 1 \end{bmatrix}, \quad \begin{bmatrix} 1 & x \\ 0 & 1 \end{bmatrix}, \quad \begin{bmatrix} 1 & y \\ 0 & 1 \end{bmatrix}.$$

Then $|G| = 2p^2$, G acts on R by conjugation, and hence $R^G = \mathbb{C}_R(G)$. To compute the latter, first note that the centralizer of $\begin{bmatrix} -1 & 0 \\ 0 & 1 \end{bmatrix}$ is precisely the set of diagonal matrices, since char $K \neq 2$. It then follows easily that $R^G = K$. But there is a natural homomorphism $S \to K$ obtained by mapping $x, y \to 0$ and this extends to a map $R \to M_2(K)$ with kernel $I \neq 0$. Since $I \triangleleft R$, I is G-invariant, but $I \cap R^G = I \cap K = 0$.

We note three properties of this example. (1) R is a prime ring but not a domain (and hence has nontrivial nilpotent elements). (2) G is inner on R. (3) R has $|G|$-torsion.

As we will see, each of these three properties is in some sense a necessary ingredient of the example. We start with a fairly easy result. In the skew group ring RG, set $\hat{G} = \sum_{x \in G} x$. Then $y\hat{G} = \hat{G}y = \hat{G}$ for $y \in G$ so $(\hat{G})^2 = |G|\hat{G}$. Also, if $r \in R$ we have

$$\hat{G}r\hat{G} = \hat{G}\sum_{x \in G} rx = \hat{G}\sum_{x \in G} xr^x = \hat{G}\sum_{x \in G} r^x.$$

Thus if we define

$$\widetilde{\text{tr}}_G(r) = \sum_{x \in G} r^x,$$

then $\widetilde{\text{tr}}_G \colon R \to R^G$ is an R^G-bimodule homomorphism and

$$\hat{G}r\hat{G} = \hat{G}\widetilde{\text{tr}}_G(r) = \widetilde{\text{tr}}_G(r)\hat{G}.$$

PROPOSITION 1. *Let G act on the ring R. Then any of the following implies the existence of fixed points in nonzero G-invariant one-sided ideals of R.*
(i) *RG is semiprime.*
(ii) *R is semiprime with no $|G|$-torsion.*
(iii) *R is prime and G is X-outer.*
(iv) *R has characteristic $p > 0$ and $|G| = p$.*

PROOF. (i) Let I be a nonzero G-invariant left ideal of R. Then $I\hat{G}$ is a left ideal of RG. Since RG is semiprime, this implies that $I\hat{G}I\hat{G} \neq 0$ so $I\widetilde{\text{tr}}_G(I)\hat{G} \neq 0$. Thus $0 \neq \widetilde{\text{tr}}_G(I) \subseteq I^G$.

(ii) and (iii) certainly imply (i).

(iv) Say $G = \langle g \rangle$ with $g^p = 1$ and let I be as above. Then g acts as a linear transformation on I and char $R = p$ so $(g-1)^p = 0$. Now note that the nilpotent transformation $g - 1$ has a nonzero kernel.

We remark that (iv) is in fact true for any finite p-group G. This covers properties (2) and (3) of the example, but what about (1)?

THEOREM 2 [2]. *Let G act on the domain R.*
(i) *If G is X-inner and faithful, then R has no $|G'|$-torsion.*
(ii) *If I is a nonzero G-invariant one-sided ideal of R, then $I^G \neq 0$.*

PROOF. (i) Say char $R = p > 0$ and let

$$H = \{q \mid q \text{ is a unit of } Q_s(R) \text{ which acts like some } g \in G \text{ on } R\}.$$

Then H is a multiplicative group which contains $C^\bullet$, the set of nonzero elements of the extended centroid C of R. Since G is both X-inner and faithful we have $H/C^\bullet \simeq G$. Therefore H is center-by-finite and this implies that the commutator subgroup H' of H is finite. Now let $q \in H'$ and suppose $q^p = 1$. Then $(q-1)^p = 0$ so $q = 1$ since $Q_s(R)$ is a domain. It follows that $p \nmid |H'|$ and hence $p \nmid |G'|$ since H' maps onto G'.

(ii) Since a subring of a domain is a domain, this easily reduces to the situation where G is simple and acts faithfully on R. Since $G_{\text{inn}} \triangleleft G$, there are three cases to consider:

    a. G is X-outer.

    b. G is X-inner and abelian.

    c. G is X-inner and nonabelian.

Case (a) is already done and for (b) we must have $|G| = p$ for some prime p. Here the result follows if either char $R = p$ or not. Finally, for (c) we have $G = G'$ so (i) implies that R has no $|G|$-torsion and again fixed points exist.

Kharchenko in fact proved the stronger result: If R has no nilpotent elements, then fixed points exist. Thus we are able to finesse $|G|$-torsion in this case. Now what can we get with the additional assumption that R has no $|G|$-torsion? We actually get the premier result on this question. First note that the augmentation map $\rho \colon RG \to R$ is given by $\rho(\sum_g r_g g) = \sum_g r_g$. For $\alpha \in RG, x \in G$ and $r \in R$ we have $\rho(\alpha x) = \rho(\alpha), \rho(x^{-1}\alpha) = \rho(\alpha)^x$ and $\rho(r\alpha) = r\rho(\alpha)$.

THEOREM 3 [1]. *Let G act on R and let I be a G-invariant one-sided ideal of R with no $|G|$-torsion. Then either $I^G \neq 0$ or I is nilpotent of bounded degree (depending on $|G|$).*

PROOF (SKETCH). Assume the I is a left ideal, form RG and let

$$IG = \left\{ \sum r_g g \in RG \mid r_g \in I \text{ for all } g \in G \right\}.$$

Then IG is a left ideal of RG and we set

$$A = \{\alpha \in IG \mid \rho(\alpha IG) = \rho(\alpha I) = 0\}.$$

We have $(RG)A \subseteq A$, $A(IG) \subseteq A$ and $A \subseteq IG$ so $\rho(A^2) = 0$. The goal is to show that $\rho(A)$ is nilpotent of bounded degree.

Define A_i to be the linear span of all elements of A of support size at most i. Then $A_0 = 0$ so $\rho(A_0) = 0$. Thus it suffices to show that $\rho(A_i)^{k_i} \subseteq \rho(A_{i-1})$ for all i and suitable integers k_i. For example, suppose $\alpha, \beta \in A$ both have the same support X with $|X| = i \geq 1$ and write $\alpha = \sum_{x \in X} a_x x$, $\beta = \sum_{x \in X} b_x x$. For each $x \in X$ form $\gamma_x = a_x \beta - \alpha b_x^x$. Then $\gamma_x \in A$ and $\mathrm{Supp}\, \gamma_x \subseteq X - \{x\}$ so $\gamma_x \in A_{i-1}$. Since $\alpha \in A$ and $b_x^x \in I$ we have $\rho(\gamma_x) = a_x \rho(\beta)$ so $a_x \rho(\beta) \in \rho(A_{i-1})$. Adding over all $x \in X$ then yields $\rho(\alpha)\rho(\beta) \in \rho(A_{i-1})$. It now follows from a pigeon hole argument that $\rho(A_i)^{k_i} \subseteq \rho(A_{i-1})$ with $k_i = \binom{|G|}{i} + 1$. Thus we have proved that $\rho(A)^n = 0$ with $n = k_1 k_2 \cdots k_{|G|}$.

Finally, suppose that $I^G = 0$ so that $\widetilde{\mathrm{tr}}_G(I) = 0$. We claim that $I\hat{G} \subseteq A$. Indeed, if $a, b \in I$, then $a\hat{G}b = \sum_x ab^x x^{-1}$ so $\rho(a\hat{G}b) = a\widetilde{\mathrm{tr}}_G(b) = 0$. Using $\rho(A)^n = 0$ and $\rho(I\hat{G}) = I \cdot |G|$ we conclude that $I^n \cdot |G|^n = 0$ and hence that $I^n = 0$ since I has no $|G|$-torsion.

It is an open problem to obtain the sharp bound for the degree of nilpotence of I in the above. It is conjectured that this bound should be $|G|$. We remark that less computational proofs of Theorem 3 can be given following the argument of

Proposition 1 or via the completely different approach of [6]. In either case, one proves that I is nilpotent but obtains no bound at all for the nilpotence degree. We briefly discuss the proof in [6]. To start with it requires

LEMMA 4. *Let $I \triangleleft R$ and let L_1 and L_2 be right ideals of R with $I + L_1 = I + L_2 = L_1 + L_2 = R$. Then $I + (L_1 \cap L_2) = R$.*

PROOF. First $I = RI = (L_1 + L_2)I$ so $I = (I \cap L_1) + (I \cap L_2)$. Next, since $I + L_1 = R$ we see that $L_1 + (I \cap L_2) = R$ and hence $L_2 = (L_1 \cap L_2) + (I \cap L_2)$. Finally, using $I + L_2 = R$ and the above we obtain $I + (L_1 \cap L_2) = R$.

LEMMA 5. *Let G act on R and let I be a G-invariant ideal of R with $I^G = 0$. Then $|G| \cdot I \subseteq JR$.*

PROOF. Let M be a maximal right ideal of R. We show that either $I \subseteq M$ or $|G| \in M$. To this end, suppose that $I \not\subseteq M$. Then $I + M = R$ and hence, for all $g \in G$, $I + M^g = R$. The previous lemma now shows that $I + \overline{M} = R$ where $\overline{M}$ is the G-invariant right ideal $\overline{M} = \bigcap_g M^g$. Thus there exists $i \in I$ with $1 - i \in \overline{M}$. Furthermore for all $g \in G$, $1 - i^g \in \overline{M}$, so summing over g yields $|G| - \widetilde{\operatorname{tr}}_G(i) \in \overline{M}$. But $\widetilde{\operatorname{tr}}_G(i) = 0$ so $|G| \in \overline{M} \subseteq M$. We conclude that $|G| \cdot I$ is contained in all such M.

Here is a sketch of the remainder of the argument. Form a suitable polynomial power series ring over R. Then G acts on $R[[X]]$ and $I[[X]]$ is a G-invariant ideal. Furthermore, if $I^G = 0$, then $I[[X]]^G = 0$ and $|G| \cdot I[[X]] \subseteq J(R[[X]])$. But $J(R[[X]])$ is nilpotent, if this extension is chosen properly, so $|G| \cdot I$ is nilpotent and hence so is I.

A natural generalization of the existence of fixed points is integrality. Namely, we ask whether I is integral in some reasonable sense over I^G. We will discuss this question in the final section. Now we move on to other skew group ring applications. For these we need $|G|^{-1} \in R$. With this assumption we can set $e = |G|^{-1} \hat{G} \in RG$. It follows easily that e is an idempotent and that

$$eRGe = eR^G = R^G e \simeq R^G.$$

This then yields an alternate route from R to its subring R^G. Namely, we can go from R to RG and then from RG to $eRGe \simeq R^G$. Since both of these steps are fairly well understood, we are able to obtain nice results in this roundabout fashion. We will discuss two of these applications below.

The first one concerns the restriction of R-modules to R^G. If V_R is a right R-module, let $\mathcal{L}(V_R)$ denote the lattice of its R-submodules. Given V we can also form the induced RG-module $W = V \otimes_R RG$ and then We is a module for $eRGe \simeq R^G$.

THEOREM 6 [3]. *Let G act on the ring R with $|G|^{-1} \in R$ and let V_R be a right R-module. If $W = V \otimes_R RG$ is the induced RG-module, then there exist order preserving maps*

$$\sigma \colon \mathcal{L}(V_{R^G}) \to \mathcal{L}(W_{RG}), \qquad \tau \colon \mathcal{L}(W_{RG}) \to \mathcal{L}(V_{R^G})$$

such that

(i) $\sigma\tau = 1$ *so σ is one-to-one and τ is onto, and*

(ii) τ *preserves direct sums.*

PROOF. Set $S = RG$ and define $\sigma\colon \mathcal{L}(W_{e_S e}) \to \mathcal{L}(W_S)$ and τ in the other direction by $A^\sigma = AS$ and $B^\tau = Be$. It is a standard, easy result that (i) and (ii) are satisfied. Thus it suffices to show that $W e_S e \simeq V_{R^G}$. But this is clear since $W = \sum_{x \in G} V \otimes x$ implies that the map $v \to \sum_x v \otimes x$ gives the necessary R^G-isomorphism.

We can now apply known results on induced modules, including Maschke's theorem, to obtain numerous consequences. We list a sample of these. Part (iv) uses the fact that this machinery also applies to bimodules.

COROLLARY 7. *Let G act on R with $|G|^{-1} \in R$ and let V_R be a right R-module.*

(i) *If V_R is Noetherian or Artinian, then so is V_{R^G}.*

(ii) *If R is right Noetherian, then R is a finitely generated R^G-module and R^G is right Noetherian.*

(iii) *If V_R is completely reducible of composition length n, then V_{R^G} is completely reducible of composition length $\leq n \cdot |G|$.*

(iv) *If R is a direct sum of n simple rings, then R^G is a direct sum of at most $n \cdot |G|$ simple rings.*

Part (ii) above is not true in general without the assumption that $|G|^{-1} \in R$. But the counterexample is not a group ring. Therefore we pose the following problem. Let H be a polycyclic-by-finite group and let G be a finite group of automorphisms of H. Then G acts on the Noetherian group rings $K[H]$ and $Z[H]$. Are the fixed subrings necessarily Noetherian?

Here is how to construct a counterexample. Let $N \triangleleft H$ with $\overline{H} = H/N$ infinite cyclic and let $G = Z_p$ act on H normalizing N. Assume that $\mathbf{C}_H(G) \subseteq N$ and that G acts trivially on $\overline{H}$. Now the fixed points of G on $Z[H]$ are spanned by the G-class sums of H and, since $|G| = p$, these class sums have size 1 or p. Suppose $h \in H \backslash N$. Then the class sum of h has size p since $\mathbf{C}_H(G) \subseteq N$ and all G-conjugates of h are congruent mod N since G acts trivially on $\overline{H}$. Thus, under the homomorphism $Z[H] \to Z[\overline{H}]$ we see that $Z[H]^G$ maps onto $R = Z + pZ[\overline{H}]$. But R is not a Noetherian ring since $pZ[\overline{H}]/p^2 Z[\overline{H}]$ is a quotient of ideals of R which has infinite R-rank.

Fortunately, such automorphisms of polycyclic-by-finite groups do not exist. Indeed we have

LEMMA 8. *Let H be a polycyclic-by-finite group and let G be a finite group of automorphisms of H. Then $[H, G] \cdot \mathbf{C}_H(G)$ has finite index in H.*

PROOF. Set $W = H \rtimes G$ so that W is also polycyclic-by-finite. Now it is a standard group theoretic fact that $[H, G] \triangleleft \langle H, G \rangle = W$ and thus we see that $N = [H, G] \cdot G \triangleleft W$. Since N is polycyclic-by-finite, it has only finitely many conjugacy classes of finite subgroups, and say these are represented by $G = G_1, G_2, \ldots, G_k$. We now apply the Frattini argument. To this end let us assume that $G_1, G_2, \ldots, G_s$ are the groups in this list which are conjugate to G in W and say $G_i = G^{w_i}$. Then if $w \in W$, we have $G^w \subseteq N$ so G^w is N-conjugate to one of these G_i and hence $G^w = G_i^x = G^{w_i x}$ for some $x \in N$. This implies that $w \in \mathsf{N}_W(G) w_i N = \mathsf{N}_W(G) \cdot Nw_i$ so we see that $W = \bigcup_1^s \mathsf{N}_W(G) \cdot Nw_i$. Hence $|W : \mathsf{N}_W(G) \cdot N| \leq s < \infty$ and the result now follows quite easily.

The final and perhaps most successful skew group ring application to Galois theory concerns the behavior of prime ideals. It is based on the earlier study of primes in RG along with the following classical fact.

LEMMA 9. *Let $e \neq 0$ be an idempotent in the ring R and let φ map the ideals of R to those of eRe by $I^\varphi = eIe$. Then φ determines a one-to-one correspondence between the primes of R not containing e and the primes of eRe. Furthermore, if P_1 and P_2 are primes of R not containing e, then $P_1 \subseteq P_2$ if and only if $P_1^\varphi \subseteq P_2^\varphi$.*

We remark that if $\tilde{P}$ is a prime ideal of eRe, then its correspondent in R is the unique largest ideal P of R with $ePe = \tilde{P}$.

Now suppose that G acts on R with $|G|^{-1} \in R$. Let T be a prime ideal of R and set $A = \bigcap_{x \in G} T^x$ so that A is a G-prime ideal. Since $RG/AG \simeq (R/A)G$ and the latter ring is semiprime, it follows that $AG = P_1 \cap P_2 \cap \cdots \cap P_n$ is an intersection of $n \leq |G|$ minimal covering primes. We then get primes $P_1^\varphi, P_2^\varphi, \ldots, P_n^\varphi$ of $eRGe \simeq R^G$. This idea was originally used by Lorenz and me to compare the prime lengths of R and of R^G. It was then carried to its completion by Montgomery who obtained

THEOREM 10 [4]. *Let G act on R with $|G|^{-1} \in R$.*

(i) *If T is a prime ideal of R, then $T \cap R^G = Q_1 \cap Q_2 \cap \cdots \cap Q_k$ is an intersection of $k \leq |G|$ minimal covering primes. Thus T lies over finitely many primes of R^G.*

(ii) *If Q is a prime ideal of R^G, then there exists a prime ideal T of R, unique up to G-conjugation, such that T lies over Q.*

(iii) *The following three versions of Going Up and Going Down hold.*

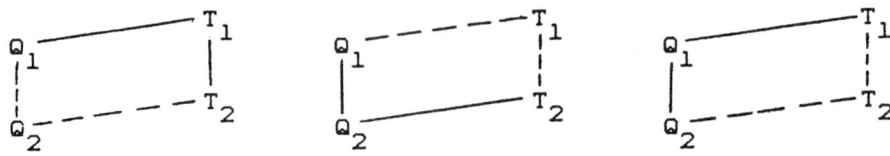

The first diagram reads as follows: Let $T_1 \supseteq T_2$ be primes of R and assume that T_1 lies over the prime Q_1 of R^G. Then there exists a prime ideal Q_2 of R^G with $Q_1 \supseteq Q_2$ and such that T_2 lies over Q_2. The other diagrams read similarly. We remark that an efficient proof of this result can now be found in [5].

We close with two relevant examples. First let $R = M_n(C)$ be the ring of $n \times n$ matrices over the complex numbers C, and for each $1 \le i \le n-1$ let d_i be the diagonal matrix $d_i = \text{diag}(1,1,\ldots,-1,1,\ldots,1)$ with -1 in the ith entry and ones elsewhere. Let G be the group of automorphisms of R generated by $g_1, g_2, \ldots, g_{n-1}$ where each g_i acts like conjugation by d_i. Then $|G| = 2^{n-1}$ and G is inner on R. The latter implies that $RG = R \otimes_C E$ where $E = C^t[G]$. But E is semiprime and is generated by the commuting elements $d_i g_i$ so $E \simeq \oplus \sum_1^{|G|} C$, a direct sum of $|G|$ copies of C. Thus $RG \simeq \oplus \sum_1^{|G|} R$ and therefore RG has $|G| = 2^{n-1}$ minimal primes. On the other hand, R^G is the subring of diagonal matrices of R so R^G has precisely n minimal primes. For $n \ge 3$ we see that there are primes lost under the φ map; that is, there are primes containing e.

Finally, let A be a simple domain over the field K and assume that A is not a division ring. Then we can choose I to be a nonzero left ideal. For example, we could take A to be the Weyl algebra $A_1(K) = K[x,y|xy - yx = 1]$ with $\text{char}\, K = 0$ and $I = Ax$. Now define $R = \begin{bmatrix} K+I & A \\ I & A \end{bmatrix} \subseteq M_2(A)$. It follows easily that R is prime so $T_2 = 0$ is a prime ideal of R. Also observe that $T_1 = \begin{bmatrix} I & A \\ I & A \end{bmatrix}$ is a maximal two-sided ideal of R with $R/T_1 = K$. Let G be the group of automorphisms of R generated by conjugation by $\text{diag}(1,-1)$. We assume of course that $\text{char}\, K \ne 2$. Then $|G| = 2$ and $R^G = \text{diag}(K+I, A)$. Thus R^G has two minimal primes one of which is $Q_2 = \text{diag}(K+I, 0)$. But $R^G/Q_2 \simeq A$ so Q_2 is also maximal. Thus we see that there exists no prime Q_1 of R^G which completes the diagram below. In other words, the missing Going Up result does indeed fail. It does not fail in RG, but the prime we get may contain e and, hence, may not correspond to a prime of R^G.

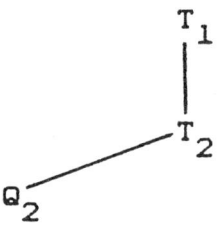

References

1. G. M. Bergman and I. M. Isaacs, *Rings with fixed-point-free group actions*, Proc. London Math. Soc. (3) **27** (1973), 69–87.
2. V. K. Kharchenko, *Generalized identities with automorphisms*, Algebra i Logika **14** (1975), 215–237; English transl. 1975, 132–148.
3. M. Lorenz and D. S. Passman, *Observations on crossed products and fixed rings*, Comm. Algebra **8** (1980), 743–779.
4. S. Montgomery, *Prime ideals in fixed rings*, Comm. Algebra **9** (1981), 423–449.
5. D. S. Passman, *It's essentially Maschke's theorem*, Rocky Mountain J. Math. **13** (1983), 37–54.
6. E. Puczylowski, *On fixed rings of automorphisms*, Proc. Amer. Math. Soc. **90** (1984), 517–518.

9. Galois Theory of Prime Rings

The Galois theory of noncommutative rings is an outgrowth of the Galois theory of fields. It was begun by E. Noether [5] in her study of inner automorphisms of central simple algebras. Later work considered division rings, complete rings of linear transformations, continuous transformation rings and Artinian rings. The best results at this time are due to Kharchenko who considered semiprime rings. We discuss this work here but restrict our attention to the prime case. The semiprime techniques are more complicated but really contain no fundamentally new ideas. The proofs invariably begin with a Zorn's lemma argument to find an idempotent maximal with some property and then proceed as in the prime case. We will follow [3] which I believe does a better job of exposition. About three quarters of that paper develops the work of Kharchenko. The last quarter is new. Also we should mention [4] which offers a fairly easy path to the simpler results in the outer case. Let us begin.

We first have to understand the allowable groups. Let R be a prime ring and let $Q = Q_s(R)$ be its symmetric Martindale ring of quotients. If the group G acts on R, we let G_0 be the subgroup of G consisting of those elements which act as X-inner automorphisms. Thus $G_0 \triangleleft G$. Furthermore, we define $B = B(G) = B_R(G)$ to be the linear span of all units q of $Q_s(R)$ such that conjugation by q induces an automorphism contained in G_0. It follows easily that B is an algebra over C, the extended centroid of R. $B(G)$ is called the algebra of the group.

We now say that G is an M-group if

(i) $|G : G_0| < \infty$,
(ii) B is a semisimple finite-dimensional C-algebra.

These are the finiteness conditions required for most results. Now let R^G denote the fixed subring of R. Since $B(G)$ is spanned by units acting like elements of G, we see that $B \subseteq C_Q(R^G)$ and thus conjugation by any unit of B will also fix R^G. This motivates the following definition. We say that G is an N-group (named for E. Noether) if it is an M-group satisfying the saturation condition

(iii) For any unit $b \in B$ we have $b^{-1}Rb = R$ and conjugation by b belongs to G.

Note that N-groups tend to be infinite since if G is not outer, then the group of units of B will usually be infinite. Note also that the above assumption on units of B is quite restrictive. One would not necessarily expect each such unit to normalize R. In fact, this part is only needed for some of the results and it would be a worthwhile task to sort out the appropriate hypothesis in each case. Of course if $B \subseteq R$, then this is not a problem. Here is an example of an N-group with $B \not\subseteq R$.

Let I denote the set of all finite matrices in $M_\infty(K)$ and let $R = K + I$ so that R is clearly a prime ring. As we pointed out earlier, $Q_s(R)$ is the set of row and column finite matrices. Fix an integer $n \geq 2$ and let B be the set of block diagonal matrices

$$B = \{\text{diag}(\alpha, \alpha, \alpha, \ldots) | \alpha \in M_n(K)\}.$$

Then $B \simeq M_n(K), B \subseteq Q_s(R)$, and any unit of B normalizes R. This shows that $G = \text{GL}_n(K)$ is an N-group with algebra of the group B.

Let R be a ring and S a subring. We set

$$\mathcal{G}(R/S) = \{\sigma \in \text{Aut } R | \sigma \text{ fixes } S\}.$$

We say that S is a Galois subring of R if S is the fixed ring of the group $\mathcal{G}(R/S)$.

THEOREM 1 [1]. *Let G be an N-group of automorphisms of the prime ring R. Then $\mathcal{G}(R/R^G) = G$.*

As we mentioned above, $C_Q(R^G) \supseteq B$. If this centralizer were larger, there might exist units in $C_Q(R^G) \setminus B$ which give rise to automorphisms of R fixing R^G. Thus, a necessary first step for the above is to prove

PROPOSITION 2. *If G is an M-group of automorphisms of R, then $C_Q(R^G) = B$.*

Suppose $R = M_2(K)$ with $K \neq GF(2)$, let $B = \begin{bmatrix} K & K \\ 0 & K \end{bmatrix} \subseteq R$ and let G be the group of units of B. Then G acts on R by conjugation and $R^G = C_R(B) = K$ so $C_Q(R^G) > B$. Of course $B = B(G)$ is not semisimple here, so the semisimplicity hypothesis is really needed.

In the outer case, the proof of Theorem 1 is fairly easy. Here G is finite so we can define $\tau(r) = \sum_{g \in G} r^g \in R^G$. If $\sigma \in \mathcal{G}(R/R^G)$, then $\tau(r)^\sigma = \tau(r)$ so $\sum_g r^{g\sigma} = \sum_g r^g$ is an identity which holds for all $r \in R$. Now use the linear independence of automorphisms. Of course the latter independence must be proved, but this at least gives the idea. Indeed the proof for more general groups proceeds in the same way. One first constructs suitable trace forms and then obtains linear independence results.

To construct trace forms we set $B^* = \text{Hom}_C(B, C)$. Then B^* is a right B-module with module structure described below. In the following, we also use $*$ to denote a dual basis.

LEMMA 3. Let $\theta\colon B^* \to B$ be a right B-module homomorphism, let $\{b_1, b_2, \ldots, b_n\}$ be a C-basis for $B = B(G)$ and let Λ be a transversal for G_0 in G. Then
$$\tau(x) = \sum_{i=1}^{n} \sum_{g \in \Lambda} (b_i x \theta(b_i^*))^g = \sum_{i,g} a_{ig} x^g b_{ig}$$
satisfies $a_{ig}, b_{ig} \in B$ and $\tau(Q) \subseteq Q^G$.

PROOF. We just look at the inner case so that $\tau(x) = \sum_i b_i x \theta(b_i^*)$. If $b \in B$, write $bb_i = \sum_j b_j c_{ij}$ with $c_{ij} \in C$. Then by definition of the B-module structure on B^*, we have $b_j^* b = \sum_i c_{ij} b_i^*$. Since $C = \mathbb{Z}(Q)$, we see that for all $x \in Q$
$$b\tau(x) = \sum_i bb_i x \theta(b_i^*) = \sum_{i,j} b_j c_{ij} x \theta(b_i^*)$$
$$= \sum_j b_j x \, \theta\left[\sum_i c_{ij} b_i^*\right] = \sum_j b_j x \theta(b_j^* b).$$

Now use the fact that θ is a B-homomorphism to conclude that $b\tau(x) = \tau(x)b$ and hence that $\tau(Q) \subseteq \mathbb{C}_Q(B)$.

Note that if B is semisimple, then $B \simeq B^*$ as right B-modules. Thus many such isomorphisms θ exist.

The next question concerns the linear independence of automorphisms. In general this is proved by truncation. Let $T(x) = \sum_i a_i x^{\sigma_i} b_i$ be a trace form as above. If $r, s \in R$, then $rT(sx) = \sum_i (r a_i s^{\sigma_i}) x^{\sigma_i} b_i$ is also a trace form and hence so is
$$\tilde{T}(x) = \sum_k r_k T(s_k x) = \sum_i \left[\sum_k r_k a_i s_k^{\sigma_i}\right] x^{\sigma_i} b_i.$$

This is called a left truncation of T. Similarly, one can truncate on the right. By studying this process one obtains

LEMMA 4. Let $T(x) = \sum_i a_i x^{\sigma_i} b_i$ be an outer trace form; that is, if σ_i is X-inner, then $\sigma_i = 1$. Assume that $\sigma_0 = 1$. Suppose that either $b_0 \neq 0$ and $\{a_i | \sigma_i = 1\}$ is C-linearly independent, or $a_0 \neq 0$ and $\{b_i | \sigma_i = 1\}$ is C-linearly independent. If I is a nonzero ideal of R, then $T(I) \neq 0$.

As a consequence we have

PROPOSITION 5. Let G be an M-group of automorphisms of the prime ring R. If I is a nonzero ideal of R, then $I \cap R^G \neq 0$.

Now we consider intermediate rings. The results here are somewhat tedious to state but there are useful special cases of interest. Let G be an N-group of automorphisms of R and let $R \supseteq S \supseteq R^G$. The problem is to decide when $S = R^H$ for some H, an N-subgroup of G. Since $\mathbb{C}_Q(R^G) = B = B(G)$ we have $\mathbb{C}_Q(S) = \mathbb{C}_B(S)$ and hence S must satisfy

THE GALOIS CENTRALIZER CONDITION [GZ]. $Z = C_B(S)$ *is a semisimple algebra spanned by its units.*

Note that Z will automatically be spanned by its units unless $C = GF(2)$ and $C \oplus C$ is a direct summand of Z. The above condition is clearly satisfied if G is outer since $B = C$ in that case. Also if R is a domain, then so is Q. Therefore B is a finite-dimensional C-division ring and hence so is Z. Next we have

THE GALOIS IDEMPOTENT CONDITION [GI]. *If e is an idempotent of B with $eS(1-e) = 0$, then there exists an idempotent $f \in Z$ with $Be = Bf$.*

Note that $Be = Bf$ implies, by taking right annihilators, that $(1-e)B = (1-f)B$ and hence we have

$$BeS(1-e)B = BfS(1-f)B = 0$$

since $fS = Sf$. Again this condition is trivially satisfied if G is outer or if R is a domain. In some sense, [GI] is used to deal with certain short identities satisfied by S; for example, $es = ese$ for all $s \in S$. Another type which must be handled is

THE GALOIS HOMOGENEITY CONDITION [GH]. *Let $b \in B \setminus 0$ and $g \in G$. Assume $bs = s^g b$ for all $s \in S$. Then $g = hg_0$ where $h \in \mathcal{G}(R/S)$ and $g_0 \in G_0$, the subgroup of X-inner automorphisms in G.*

Suppose that b is a unit of B in the above. Then since G is an N-group, conjugation by b yields an automorphism g_0^{-1} of R with $g_0 \in G_0$, and then clearly $gg_0^{-1} = h \in \mathcal{G}(R/S)$. Thus [GH] is automatically satisfied if G is outer, R is a domain, or if G is inner. For the latter we merely take $g_0 = g$ and $h = 1$. Finally

THE GALOIS CANCELLATION CONDITION [GC]. *Let $I \triangleleft S$ with $\underline{r}_R(I) = 0$ and let $r \in R$ with $Ir \subseteq S$. Then $r \in S$.*

Suppose that I, r and S are as above and assume in addition that $S = R^H$. If $h \in H$, then $I, Ir \subseteq R^H$ imply easily that $I(r - r^h) = 0$ and hence that $r = r^h$. Thus $r \in R^H = S$. The main result on intermediate rings is

THEOREM 6 [1]. *Let G be an N-group of automorphisms of the prime ring R and let $R \supseteq S \supseteq R^G$ so that S is an intermediate ring. Then $S = R^H$ for some N-subgroup H of G if and only if S satisfies [GZ], [GI], [GH] and [GC].*

In view of our earlier comments, there are a number of special cases where this simplifies—for example, if G is inner, or if R is a domain or

COROLLARY 7 [1]. *Let G be a finite X-outer group of automorphisms of the prime ring R and let $R \supseteq S \supseteq R^G$ so that S is an intermediate ring. Then $S = R^H$ for some subgroup H of G if and only if S satisfies [GC] (such S are called anti-ideals).*

In the proof of Theorem 6 one actually shows

PROPOSITION 8. *Let G be an N-group and let $R \supseteq S \supseteq R^G$ with S satisfying [GZ], [GI] and [GH]. If $H = \mathcal{G}(R/S)$, then there exists $I \subseteq S \subseteq R^H$ with $I \triangleleft R^H$ and $\underline{r}_R(I) = 0$.*

Three comments are now in order. The first is an example of Teichmüller. Let K be a field, σ a field automorphism of finite order at least 3, and let $k = K^{\langle \sigma \rangle}$ be the fixed field. Set $R = M_2(K)$ and let $G = \langle \mathrm{GL}_2(K), \sigma \rangle$ act on R. Then G is an N-group of automorphisms with $B = M_2(K)$ and $R^G = k$. Now set $S = \{\mathrm{diag}(a, a^\sigma) | a \in K\}$. Then $R \supseteq S \supseteq R^G$, $Z = C_B(S)$ is the set of diagonal matrices of R, and $S \simeq K$. It follows from this that S satisfies [GZ] and [GC]. For [GI], let e be an idempotent of R with $eS(1-e) = 0$. Then $eZ(1-e) = 0$ since Z is generated by K and S. But Z is a fixed ring with $C_B(Z) = Z$ so we can use the fact that Z satisfies [GI]. Finally, since $o(\sigma) \geq 3$ it follows easily that if $H = \mathcal{G}(R/S)$ then $H \subseteq \mathrm{GL}_2(K)$ is inner, and hence $R^H = Z > S$. In other words, S is not a fixed ring and indeed S does not satisfy [GH] since if $b = \begin{pmatrix} 0 & 1 \\ 0 & 0 \end{pmatrix} \in B$ then $bs = s^\sigma b$ for all $s \in S$ but $\sigma \notin HG_0 = \mathrm{GL}_2(K)$.

The second comment is that [GI] is essentially equivalent to the bimodule property where the latter is a key ingredient in the overall proof. Its definition is contained in the following statement.

LEMMA 9. *Let G be an M-group and let $R \supseteq S \supseteq R^G$ with S satisfying [GI]. Let M be an (R, S)-subbimodule of Q. Then there exists an idempotent $f \in Z = C_Q(S)$ with $\underline{r}_Z(M) = (1-f)Z$ such that $M \supseteq If$ for some nonzero ideal I of R.*

PROOF (SKETCH). Step 1. R^G has the bimodule property. Let $e \in B = B(G)$ be an idempotent with $Me \neq 0$ and say $m \in M$ with $me \neq 0$. Choose a trace form $\tau(x) = \sum a_{ig} x^g b_{ig}$ and a nonzero ideal I of R with $0 \neq \tau(I) \subseteq R^G$, $a_{11} = e$, and $\{b_{i1}\}$ appropriately related to e. If $T(x) = m\tau(x)$, then $T \neq 0$ and $T(I) \subseteq M$ since $mR^G \subseteq M$. Furthermore, if $\tilde{T} = \sum r_k T(s_k x)$ is any truncation of T, then since $RM \subseteq M$ we have $\tilde{T}(I) \subseteq M$. Truncating as far as possible, we obtain $\tilde{T}(x) = ax\beta$ with $a \in R\backslash 0$, $\beta \in B$ and $\beta e \neq 0$ by the choice of $\{b_{i1}\}$. Thus $(RaI)\beta \subseteq M$ and $\beta e \neq 0$. Now vary the e's appropriately.

Step 2. S has the bimodule property. Since $S \supseteq R^G$, M is an (R, R^G)-bimodule so there exists $e \in B$ and $0 \neq I \triangleleft R$ with $M \supseteq Ie$ and $M(1-e) = 0$. But M is a right S-module so $M \supseteq IeS$ and hence $IeS(1-e) = 0$. Thus $eS(1-e) = 0$ and now switch e to $f \in Z$ using [GI].

Our third comment concerns a strengthening of the N-group assumption. We say that G is an F-group on R if it is an N-group with $B(G)$ simple. This motivates the replacement of [GZ] by

[GZ']. $Z = C_B(S)$ *is a simple subalgebra of B.*

It turns out that [GZ'] implies [GH]. (Notice that S did not satisfy [GZ'] in the Teichmüller example.) The following is a sample of the various correspondence theorems which one obtains by considering special cases.

COROLLARY 10. *Let G be an N-group on the prime ring R. Then the maps*

$$H \to R^H, \qquad S \to \mathcal{G}(R/S)$$

yield a one-to-one correspondence between the F-subgroups H of G and the subrings S with $R \supseteq S \supseteq R^G$ which satisfy [GZ'], [GI] and [GC].

Let us move on. Let G be an N-group on R. Then it is not true that R^G is prime, but it is semiprime. Note that G acts on R, so it acts on Q and then on B. Thus G permutes the centrally primitive idempotents of B and we call the sum of the idempotents in each orbit the G-centrally primitive idempotents of B. These determine the minimal primes of R^G. More generally we have

PROPOSITION 11. *Let G be an N-group and let $R \supseteq S \supseteq R^G$ with S satisfying [GZ], [GI] and [GH]. Set $Z = \mathsf{C}_B(S)$, $H = \mathcal{G}(R/S)$ and let $f_1, f_2, \ldots, f_n$ be the H-centrally primitive idempotents of Z. Define $P_i \triangleleft S$ by $P_i = \operatorname{ann}_S f_i$.*
  (i) *$P_1, P_2, \ldots, P_n$ are the distinct minimal primes of S and $\bigcap_1^n P_i = 0$.*
  (ii) *$\operatorname{ann}_Z P_i = Z f_i$.*
  (iii) *If f is any nonzero idempotent in $Z f_i$, then $\operatorname{ann}_S f = P_i$ and the map $s \to fs$ yields an isomorphism $S/P_i \simeq fS$.*

The above of course applies with $H = G, S = R^G$ and $Z = B$. Recall from the Galois theory of fields that if $K \supseteq F_1, F_2 \supseteq K^G$, then any K^G-isomorphism $F_1 \to F_2$ extends to an automorphism of K. There is an analogous but more complicated result for prime rings.

THEOREM 12 [**2, 3**]. *Let G be an N-group of automorphisms of the prime ring R and let $R \supseteq S, \overline{S} \supseteq R^G$ with both S and $\overline{S}$ satisfying [GZ], [GI] and [GH]. Given $\varphi \colon S \to \overline{S}$, an isomorphism which is the identity on R^G, let P be a minimal prime of S and let $\overline{P} = P^\varphi$ be the corresponding minimal prime of $\overline{S}$. Then there exists $g \in G$ which "induces" the isomorphism $\varphi \colon S/P \to \overline{S}/\overline{P}$.*

More precisely, set $Z = \mathsf{C}_B(S), \overline{Z} = \mathsf{C}_B(\overline{S})$ and let $e, \overline{e}$ be primitive idempotents of Z and $\overline{Z}$ which annihilate P and $\overline{P}$ respectively. Then there exists $g \in G$ with $(es)^g = \overline{e}s^\varphi$ for all $s \in S$. Now use $S/P \simeq eS$ and $\overline{S}/\overline{P} \simeq \overline{e}\overline{S}$. This works best when Z and $\overline{Z}$ are both simple.

COROLLARY 13. *Let G be an N-group on R and let $R \supseteq S, \overline{S} \supseteq R^G$ with both S and $\overline{S}$ satisfying [GZ'] and [GI]. If $\varphi \colon S \to \overline{S}$ is an isomorphism which is the identity on R^G, then φ is the restriction of some $g \in G$.*

Here are some examples. First let $R = M_4(K)$ and let $G = \operatorname{GL}_4(K)$. Set

$$S = \{\operatorname{diag}(a,a,a,b) | a, b \in K\}, \qquad \overline{S} = \{\operatorname{diag}(a,a,b,b) | a, b \in K\}.$$

Then S and $\overline{S}$ are fixed rings of subgroups of G and $S \simeq \overline{S}$, but this isomorphism is not extendable to all of R since $Z \not\simeq \overline{Z}$.

Now let $R = M_2(K)$ and let $\sigma \neq 1$ for an automorphism of K of finite order. If $G = \langle \operatorname{GL}_2(K), \sigma \rangle$, then $R^G = k = K^{\langle \sigma \rangle}$. Set $S = \{\operatorname{diag}(a,b) | a, b \in K\}$, again the fixed ring of a subgroup of G, and let $\varphi\colon S \to S$ be given by $\mathrm{diag}(a,b) \to \mathrm{diag}(a^\sigma, b)$. In this case, φ does extend to $Z = S$ but again not to R. Rather it is the restriction of two different elements of G, one for each prime factor of S.

Finally, there exist examples to show that the hypothesis of Corollary 13 cannot be weakened to the more natural condition that Z and $\overline{Z}$ are H- and $\overline{H}$-simple, respectively. They must be simple.

In closing we discuss normal subgroups. Let G be an N-group of automorphisms and let W be an M-subgroup. Then W can be completed to an N-subgroup $\tilde{W}$ of G by adding to W the action of all units of $B(W)$. Clearly $B(W) = B(\tilde{W})$ and $R^W = R^{\tilde{W}}$. Let $H \subseteq G$. We say that H is almost normal in G if $W = \mathsf{N}_G(H)$ is an M-group with $\tilde{W} = G$. Since automorphisms only extend in a reasonable manner from subrings S with Z simple, we must now restrict our attention to F-subgroups.

THEOREM 14 [3]. *Let G be an N-group of automorphisms of the prime ring R and let H be an F-subgroup of G. Then R^H is N-group Galois over R^G if and only if H is almost normal in G.*

We note in the above that R^H is prime and that N-group Galois means that R^G is the fixed ring of the N-group $\mathcal{G}(R^H/R^G)$. Furthermore, by Corollary 13, if H is any F-subgroup of G, then any automorphism of R^H fixing R^G is the restriction of some $g \in G$. Indeed g must belong to $\mathsf{N}_G(H) = W$ so $\mathcal{G}(R^H/R^G) \simeq W/H$.

Thus we see that the Galois theory of prime rings (and also of semiprime rings) is a highly developed subject which parallels the Galois theory of fields. What is missing, however, are applications of this theory. It would be nice to find intrinsic applications of this machinery and we pose this as a serious problem to be considered.

References

1. V. K. Kharchenko, *Galois theory of semiprime rings*, Algebra i Logika **16** (1977), 313–363; English transl. 1978, 208–258.
2. ____, *Algebras of invariants of free algebras*, Algebra i Logika **17** (1978), 478–487; English transl. 1979, 316–321.
3. S. Montgomery and D. S. Passman, *Galois theory of prime rings*, J. Pure Appl. Algebra **31** (1984), 139–184.
4. ____, *Outer Galois theory of prime rings*, Rocky Mountain J. Math. **14** (1984), 305–318.
5. E. Noether, *Nichtkommutative algebra*, Math. Z. **37** (1933), 514–541.

10. Rings and Fixed Rings

In this last section we return to the action of a finite group G on a ring R. We are naturally interested in the relationship between R and the fixed ring R^G. For example, if R has a certain property, we ask whether it is inherited by R^G and vice versa. We will consider this at least for properties related to chain conditions. A wider view of the subject can be found in the survey [6] and the monograph [10].

The first question of interest is actually of a different nature. We ask whether R is integral in some sense over R^G. To see what is involved here, we begin with a special case. Let S be an algebra over the field K and set $R = M_n(S)$. Furthermore, let G be a finite absolutely irreducible subgroup of $\mathrm{GL}_n(K)$ so that the K-linear span of G is $M_n(K)$. For example, if $\mathrm{char}\, K = 0$, then Sym_{n+1} has such an irreducible representation. Then G acts on R and $R^G = S$, embedded as scalars. Thus at the very least we expect $M_n(S)$ to be integral over S. Fortunately this is a result of the extremely clever paper [11].

We start with some definitions. Let R be a ring (possibly without 1) and let T be a subring. If $r_1, r_2, \ldots, r_m \in R$, then a T-monomial is a product of these r_i's in some order and elements of T with at least one element of T occurring. For example, if $t_1, t_2 \in T$, then $r_1^2 t_1 r_2 t_2 r_1 r_3$ is a T-monomial but $r_1^2 r_2 r_1 r_3$ is not (unless $1 \in T$). The degree of such a monomial is the total degree in all the r_i's. We say that R is fully integral over T of degree m if for any $r_1, r_2, \ldots, r_m \in R$ we have
$$r_1 r_2 \cdots r_m = \varphi(r_1, r_2, \ldots, r_m)$$
where φ is a sum of T-monomials in the r_i's of degree less than m. In particular, setting $r_1 = r_2 = \cdots = r_m = r \in R$ we see that R is therefore Schelter integral over T.

If A is a ring with 1, we let $\{e_{ij}\}$ denote the set of matrix units of $M_n(A)$.

THEOREM 1 [11]. *Let A be a ring with 1 and let $R \supseteq T$ be subrings of $M_n(A)$ (possibly without 1). Assume that*
 (i) $[\sum_1^k e_{ii}] R [\sum_1^k e_{ii}] \subseteq R$ *for all* $k = 1, 2, \ldots, n$.
 (ii) T *consists of diagonal matrices and* $e_{ii} R e_{ii} = e_{ii} T e_{ii}$ *for all* $i = 1, 2, \ldots, n$.
Then there exists an integer $m = m(n) \geq 1$ such that R is fully integral over T of degree m.

PROOF (SKETCH). For each $k = 1, 2, \ldots, n$, we embed $M_k(A)$ into $M_n(A)$ as the $k \times k$ upper left corner. The goal is to show by induction on k that $R \cap M_k(A)$ is fully integral over T of degree $m(k)$. For convenience we just consider Schelter integrality.

First let $k = 1$ and choose $r = \begin{bmatrix} a & 0 \\ 0 & 0 \end{bmatrix} \in R \cap M_1(A)$. Then (ii) implies that there exists $t = \begin{bmatrix} a & 0 \\ 0 & * \end{bmatrix} \in T$. Thus $r^2 = tr$ and $m(1) = 2$.

Now let $k > 1$ and for any $r \in R \cap M_k(A)$ write

$$r = \begin{bmatrix} r' & r'' & 0 \\ * & * & 0 \\ 0 & 0 & 0 \end{bmatrix}$$

where r' is the $(k-1) \times (k-1)$ corner block. Furthermore, for this r set

$$\tilde{r} = \begin{bmatrix} r' & 0 & 0 \\ 0 & 0 & 0 \\ 0 & 0 & 0 \end{bmatrix}$$

so that $\tilde{r} \in R$ by assumption (i). Given $r_1, r_2 \in R \cap M_k(A)$ we say $r_1 \equiv r_2$ if and only if $r_1'' = r_2''$. We note two properties of $\equiv$.

(1) Let $r_1, r_2, r \in R \cap M_k(A)$ and $t \in T$. If $r_1 \equiv r_2$, then $\tilde{r} r_1 \equiv \tilde{r} r_2$ and $tr_1 \equiv tr_2$.

(2) Let $r_1, r \in R \cap M_k(A)$. Then there exists $t \in T$ with $\tilde{r}_1 r \equiv r_1 r - r_1 t$. Indeed there exists $t \in T$ with

$$r - t = \begin{bmatrix} * & r'' & 0 \\ * & 0 & 0 \\ 0 & 0 & * \end{bmatrix}$$

and then

$$r_1(r - t) = \begin{bmatrix} * & r_1' r'' & 0 \\ * & * & 0 \\ 0 & 0 & 0 \end{bmatrix} \equiv \tilde{r}_1 r.$$

Suppose $k = 2$ and let $r \in R \cap M_2(K)$ be given. Then the $k = 1$ result implies that $\tilde{r}^2 - t_1 \tilde{r} = 0$ for some $t_1 \in T$ so $(\tilde{r}^2 - t_1 \tilde{r})r \equiv 0$. Using (1) and (2) we have

$$\tilde{r}^2 r \equiv \tilde{r}(r^2 - rt_2) \equiv r^3 - r^2 t_2 - rt_3, \qquad t_1 \tilde{r} r \equiv t_1(r^2 - rt_4)$$

so

$$s = r^3 - r^2 t_2 - rt_3 - t_1 r^2 + t_1 rt_4 \equiv \tilde{r}^2 r - t_1 \tilde{r} r \equiv 0.$$

Thus s looks like

$$s = \begin{bmatrix} * & 0 & 0 \\ * & * & 0 \\ 0 & 0 & 0 \end{bmatrix}$$

and it is now easy to find a monic polynomial satisfied by s. This completes the proof for $k = 2$. With a little more care, this argument can be made to hold for all $k > 1$.

We note that $m(n) \leq 2^{2^{n+1}}$ and we use this function in the corollaries below. Let R be a ring with 1 and let $1 = e_1 + e_2 + \cdots + e_n$ be a decomposition of 1 into orthogonal idempotents. If $\mathcal{E} = \{e_1, e_2, \ldots, e_n\}$ then we have $C_R(\mathcal{E}) = e_1 R e_1 + e_2 R e_2 + \cdots + e_n R e_n$. We say that a subring T of $C_R(\mathcal{E}) = S$ is an $\mathcal{E}$-transversal if $Se_i = Te_i$ for all i.

COROLLARY 2 [**12**]. *Let R be a ring with 1 and let $1 = e_1 + e_2 + \cdots + e_n$ be a decomposition of 1 into orthogonal idempotents. Set $\mathcal{E} = \{e_1, e_2, \ldots, e_n\}$ and let T be an $\mathcal{E}$-transversal for $C_R(\mathcal{E})$. Then R is fully integral over T of degree $m(n)$.*

PROOF. We embed R into $M_n(R)$ by mapping each $r \in R$ to $[e_i r e_j]$. This is in fact a monomorphism in which the identity elements do not correspond. Now we merely note that the image of R satisfies the truncation property (i) of Theorem 1 and that the image of T satisfies property (ii).

This proof is reminiscent of duality, in that we obtain a result about R by embedding it in $M_n(R)$. Even more to the point is the following result of Bergman (see [**12**]).

COROLLARY 3. *Let R be a G-graded ring with G finite. Then R is fully integral over the base ring $R(1)$ of degree $m(|G|)$.*

PROOF. Recall that $M_G(R) \supseteq R\{1\} = \{\alpha | \alpha_{xy} \in R(x^{-1}y)\}$. Furthermore, $R\{1\}$ is closed under truncation and contains $R(1)$ embedded as scalars. It follows from Theorem 1 that $R\{1\}$ is fully integral over $R(1)$ of degree $m(|G|)$. Now we embed R into $M_G(R)$ by mapping $r = \sum r(x)$ to the matrix α with $\alpha_{xy} = r(x^{-1}y)$. Since the image of R is contained in $R\{1\}$ and since $R(1)$ maps to the $R(1)$ scalars, the result follows.

COROLLARY 4 [**12**]. *Let G be a finite abelian group acting on the ring R (possibly without 1) and assume that $R = |G| \cdot R$. Then R is fully integral over R^G of degree $m(|G|)$.*

PROOF. It is easy to reduce this to the generic case where $|G|^{-1} \in R$ and R contains suitable $|G|$th roots of unity. Since G is abelian, it then follows that R is graded by the dual group of G and that $R(1) = R^G$. Now apply Corollary 3.

The original proof of this result used skew group rings and Corollary 2. We remark that the above integrality is not known for any nonabelian group, even $G = \text{Sym}_3$. We also note that the hypothesis $R = |G| \cdot R$ is required here. For example, let K be a field of characteristic $p > 0$, let $S = K\langle x, y \rangle$ be the free K-algebra on two generators and set $R = M_2(S)$. If G is the group of units of R generated by $\begin{bmatrix} 1 & x \\ 0 & 1 \end{bmatrix}$ and $\begin{bmatrix} 1 & y \\ 0 & 1 \end{bmatrix}$, then G is abelian, $|G| = p^2$ and

$$R^G = \left\{ \begin{bmatrix} a & b \\ 0 & a \end{bmatrix} \middle| a \in K, b \in S \right\}.$$

Certainly the scalar matrices x and y are not integral over R^G.

Now suppose that the ring R of Corollary 4 satisfies $R^G = 0$. Then since all lower degree monomials involve R^G, we see that R is nilpotent of index $m(|G|)$. This nilpotence bound is much too big ($|G|$ suffices for G abelian) but, nevertheless, Corollary 4 should be viewed as an extension of the Bergman-Isaacs theorem at least for G abelian. Obviously there is much work still to be done here when G is nonabelian.

Let us move on to consider chain conditions and related properties. We return to the assumption that $1 \in R$. We start by listing applications of the skew group ring.

PROPOSITION 5. *Let G act on the ring R and suppose that R is semiprime with no $|G|$-torsion. Let E be a right ideal of R and let A be a right ideal of R^G.*
 (i) R^G *is semiprime.*
 (ii) *If $0 \neq E$ is G-invariant, then $E^G = E \cap R^G \neq 0$.*
 (iii) $|G| \cdot A \subseteq |G| \cdot (AR \cap R^G) \subseteq A$.
 (iv) *If E ess R, then $E \cap R^G$ ess R^G.*
 (v) *If A ess R^G, then AR ess R.*

PROOF. (i) We can assume that $|G|^{-1} \in R$. Since R is semiprime, so is RG and hence, also, $eRGe \simeq R^G$ where $e = |G|^{-1}\hat{G}$. (ii) This follows from the Bergman-Isaacs theorem or the fact that RG is semiprime. (iii) Here we merely note that $\widetilde{\mathrm{tr}}_G(AR) = A\widetilde{\mathrm{tr}}_G(R) \subseteq A$ and that $\widetilde{\mathrm{tr}}_G(r) = |G| \cdot r$ if $r \in R^G$. (iv) Since G is finite, $\bigcap_g E^g$ ess R so we can assume that E is G-invariant. If A is a nonzero right ideal of R, then $E \cap AR \neq 0$ so $|G|(E \cap AR)^G \neq 0$ and hence $A \cap E^G \neq 0$. (v) This uses the essential version of Maschke's theorem. Observe that R is a right RG-module where R acts by right multiplication and G acts as the given automorphisms. Thus the RG-submodules of R are precisely the G-invariant right ideals. We claim first that AR ess_{RG} R. Indeed if E is G-invariant and $AR \cap E = 0$, then $A \cap E^G = 0$. Since A ess R^G we have $E^G = 0$ and hence $E = 0$ by (ii). Now apply Maschke's theorem to deduce that AR ess_R R.

In view of the above, ring theoretic properties described by essential right ideals should translate readily between R and R^G. For example, R is semisimple Artinian if and only if it has no proper essential right ideals. Therefore we obtain

THEOREM 6 [2, 9]. *Let G act on R where R is a semiprime ring with no $|G|$-torsion. Then R is semisimple Artinian if and only if R^G is.*

PROOF. Assume R^G is semisimple Artinian and let E ess R. Then $E \cap R^G$ ess R^G so $1 \in E \cap R^G \subseteq E$. Conversely, let R be semisimple Artinian. Since R has no $|G|$-torsion, this implies that $|G|^{-1} \in R$. If A ess R^G, then AR ess R so $1 \in AR$ and $|G| \in A$.

The semiprime Goldie property can also be characterized by essential right ideals. The following isolates the last few steps in the proof of Goldie's theorem.

LEMMA 7. *Let T be a multiplicatively closed set of regular elements of R and suppose*

(i) *$t \in T$ implies tR ess R, and*
(ii) *E ess R implies $E \cap T \neq \emptyset$.*

Then T is a right divisor set in R, RT^{-1} is semisimple Artinian and R is a semiprime Goldie ring with classical ring of quotients $Q_{cl}(R) = RT^{-1}$.

With this we can quickly prove the following result.

THEOREM 8 [1, 5, 7]. *Let G act on R where R is semiprime with no $|G|$-torsion. Then R is semiprime Goldie if and only if R^G is. When this occurs then $Q_{cl}(R) = RT^{-1}$ where T is the set of regular elements of R^G and $Q_{cl}(R)^G = Q_{cl}(R^G)$.*

PROOF. We already know that R^G is semiprime. Suppose R is Goldie. Then $R^G \subseteq R \subseteq Q_{cl}(R)$ and the latter ring is Artinian so R^G satisfies the maximal condition on right annihilators. Furthermore, if $A_1 \oplus \cdots \oplus A_k$ is a direct sum of right ideals of R^G, then Proposition 5 implies, by looking at the fixed points in a nonzero intersection, that $A_1 R \oplus \cdots \oplus A_k R$ is direct. Since the rank of R is finite we conclude that R^G is Goldie.

Conversely, suppose R^G is Goldie. Then $Q_{cl}(R^G) = R^G T^{-1}$ where T is the set of regular elements of R^G. Now T is a multiplicatively closed subset of R and it consists of regular elements since for each $t \in T$ the G-invariant one-sided ideals $\underline{r}_R(t)$ and $l_R(t)$ have no nonzero fixed points. We claim that T satisfies (i) and (ii) above. For (i), if $t \in T$, then tR^G ess R^G so $tR = (tR^G)R$ is essential in R. For (ii), if E ess R, then $E \cap R^G$ ess R^G so $(E \cap R^G) \cap T \neq \emptyset$. The lemma now implies that R is semiprime Goldie with $Q_{cl}(R) = RT^{-1}$. Since $T \subseteq R^G$ this also yields $Q_{cl}(R)^G = R^G T^{-1} = Q_{cl}(R^G)$.

We move on to the Noetherian condition. We already know that if R is right Noetherian with $|G|^{-1} \in R$, then R^G is right Noetherian and R is a finitely generated right R^G-module. Conversely, we have

THEOREM 9 [3]. *Let G act on R, a semiprime ring with no $|G|$-torsion. If R^G is right Noetherian, then so is R.*

PROOF. Since R^G is semiprime and Noetherian, it is a Goldie ring and hence, by the above, so is R. Also, $Q_{cl}(R) = RT^{-1}$ and $Q_{cl}(R^G) = R^G T^{-1}$ where T is the set of regular elements of R^G. Now $Q_{cl}(R)$ is left Noetherian (it is in fact semisimple Artinian) and it contains $|G|^{-1}$ so $Q_{cl}(R)$ is a finitely generated left module over $Q_{cl}(R)^G = R^G T^{-1}$. It follows easily that there exist $r_1, r_2, \ldots, r_n \in R$ with $Q_{cl}(R) = \sum_1^n Q_{cl}(R)^G r_i$.

Let $r \in R$ and suppose that $\widetilde{\text{tr}}_G(r_i r) = 0$ for all i. Then $\widetilde{\text{tr}}_G(Q_{cl}(R)r) = 0$ so $\widetilde{\text{tr}}_G(Rr) = 0$. Hence $\widetilde{\text{tr}}_G(\sum_g Rr^g) = 0$ and we have $r = 0$ since $\sum_g Rr^g$ is a G-invariant left ideal with no nonzero fixed points. Finally, we map R to $\oplus \sum_1^n R^G$ by sending $r \in R$ to $\oplus \sum_1^n \widetilde{\text{tr}}_G(r_i r)$. This is a right R^G-module homomorphism and we know that it is an embedding. Since R^G is right Noetherian, we conclude that R is a Noetherian R^G-module and, hence, that R is right Noetherian.

Certainly we cannot drop the assumption of no $|G|$-torsion in this theorem. Now we mention a related result of Kharchenko which comes out of his Galois theory techniques.

THEOREM 10 [8]. *Let R be a prime ring and let G be an M-group of automorphisms of R. Set*

$$I = \{r \in R | rR \subseteq \text{ a finitely generated right } R^G\text{-submodule of } R$$
$$\text{and } Rr \subseteq \text{ a finitely generated left } R^G\text{-submodule of } R\}.$$

Then I is a nonzero two-sided ideal of R.

PROOF (SKETCH). It is clear that $I \triangleleft R$. The goal is to show that $I \neq 0$. Let $T(x)$ be a trace form and let $0 \neq A \triangleleft R$ with $0 \neq T(A) \subseteq R^G$. By truncating T as far as possible, we obtain $\tilde{T}(x) = rx\beta$ with $r \in R\setminus 0$ and $\beta \in B(G)\setminus 0$. Since $\tilde{T}(x) = \sum_k r_k T(s_k x)$ we see that

$$rA\beta = \tilde{T}(A) \subseteq \sum_k r_k T(s_k A) \subseteq \sum_k r_k R^G.$$

If $\beta = 1$, then $rA \subseteq I$ (at least it satisfies the right module condition) and we are done. The idea now to vary the initial choice of T to find enough β's to span B. Then we add these appropriately to obtain a linear combination equal to 1.

This result can be applied to a finite group G provided we can guarantee that it is an M-group. Here it is nice, but certainly not necessary, to bring in skew group rings.

LEMMA 11. *Let G act on the prime ring R and set $S = Q_s(R)$. Then the augmentation map $\eta \colon \sum_x s_x x \to \sum_x s_x$ defined on $E = \mathbb{C}_{SG}(S) \subseteq SG$ is an antihomomorphism onto $B(G)$.*

PROOF. The map η is clearly additive. For multiplication, we first observe that $\sum_x s_x x \in E$ if and only if each $s_x x \in E$. Now if $sx, ty \in E$ then since t commutes with sx we have $sx \cdot ty = tsxy$ and hence

$$\eta(sx \cdot ty) = ts = \eta(ty)\eta(sx).$$

Thus η is an antihomomorphism into S. Finally, $sx \in E$ with $s \neq 0$ if and only if s is a unit of S which acts like x^{-1} on R. We conclude that $\eta(E) = B(G)$.

In particular, since $E \simeq C^t[G_{\text{inn}}]$, a twisted group algebra over the extended centroid C, we see that if G is finite and if R has no $|G_{\text{inn}}|$-torsion then E and hence $B(G)$ is semisimple. Thus G is an M-group on R. As we remarked earlier, Kharchenko's work handles semiprime rings. Here even the definition of M-group is tedious, but at least we know that if G is finite and R has no $|G|$-torsion, then G is an M-group. Thus

THEOREM 12 [8]. *Let G be a finite group acting on the ring R which is semiprime with no $|G|$-torsion. Set*

$$I = \{ r \in R | rR \subseteq \text{ a finitely generated right } R^G\text{-submodule of } R$$
$$\text{and } Rr \subseteq \text{ a finitely generated left } R^G\text{-submodule of } R \}.$$

Then I is an essential two-sided ideal of R.

Now suppose that G and R are as above and that R^G is Noetherian. Then R^G is Goldie so R is Goldie and hence the ideal I contains a regular element r. Thus $R \simeq rR$ is a finitely generated R^G-module and we have obtained an alternate proof of Theorem 9.

Finally, we briefly mention

THEOREM 13 [4]. *Let G be a finite group acting as automorphisms on the modular lattice L. Then there exists a strictly increasing map $f \colon L \to \oplus \sum L^G$ where the latter is a finite direct sum of copies of the sublattice L^G of fixed points under G.*

Hence if L^G satisfies any of a large class of chain conditions, then so does L. Clearly this has numerous applications to the relationship between rings and fixed rings.

References

1. M. Cohen, *Semiprime Goldie centralizers*, Israel J. Math. **29** (1975), 37–45; Addendum, **24** (1976), 89–93.
2. M. Cohen and S. Montgomery, *Semisimple Artinian rings of fixed points*, Canad. Math. Bull. **18** (1975), 189–190.
3. D. R. Farkas and R. L. Snider, *Noetherian fixed rings*, Pacific J. Math. **69** (1977), 347–353.
4. J. W. Fisher, *Chain conditions for modular lattices with finite group actions*, Canad. J. Math. **31** (1979), 558–564.
5. J. W. Fisher and J. Osterburg, *Semiprime ideals in rings with finite group actions*, J. Algebra **50** (1978), 488–502.
6. ____, *Finite group actions on noncommutative rings: a survey since 1970*, Ring Theory and Algebra, III (Proc. Third Conf., Univ. Oklahoma, 1979), Marcel Dekker, New York, 1980, pp. 357–393.
7. V. K. Kharchenko, *Galois extensions and quotient rings*, Algebra i Logika **13** (1974), 460–484; English transl. 1975, 264–281.
8. ____, *Galois theory of semiprime rings*, Algebra i Logika **16** (1977), 313–363; English transl. 1978, 208–258.
9. J. Levitzki, *On automorphisms of certain rings*, Ann. of Math. **36** (1935), 984–992.
10. S. Montgomery, *Fixed rings of finite automorphism groups of associative rings*, Lecture Notes in Mathematics, No. 818, Springer, Berlin, 1980.
11. R. Paré and W. Schelter, *Finite extensions are integral*, J. Algebra **53** (1978), 477–479.
12. D. S. Passman, *Fixed rings and integrality*, J. Algebra **68** (1981), 510–519.

Other Titles in This Series

88 **Craig Huneke,** Tight closure and its applications, 1996
87 **John Erik Fornæss,** Dynamics in several complex variables, 1996
86 **Sorin Popa,** Classification of subfactors and their endomorphisms, 1995
85 **Michio Jimbo and Tetsuji Miwa,** Algebraic analysis of solvable lattice models, 1994
84 **Hugh L. Montgomery,** Ten lectures on the interface between analytic number theory and harmonic analysis, 1994
83 **Carlos E. Kenig,** Harmonic analysis techniques for second order elliptic boundary value problems, 1994
82 **Susan Montgomery,** Hopf algebras and their actions on rings, 1993
81 **Steven G. Krantz,** Geometric analysis and function spaces, 1993
80 **Vaughan F. R. Jones,** Subfactors and knots, 1991
79 **Michael Frazier, Björn Jawerth, and Guido Weiss,** Littlewood-Paley theory and the study of function spaces, 1991
78 **Edward Formanek,** The polynomial identities and variants of $n \times n$ matrices, 1991
77 **Michael Christ,** Lectures on singular integral operators, 1990
76 **Klaus Schmidt,** Algebraic ideas in ergodic theory, 1990
75 **F. Thomas Farrell and L. Edwin Jones,** Classical aspherical manifolds, 1990
74 **Lawrence C. Evans,** Weak convergence methods for nonlinear partial differential equations, 1990
73 **Walter A. Strauss,** Nonlinear wave equations, 1989
72 **Peter Orlik,** Introduction to arrangements, 1989
71 **Harry Dym,** J contractive matrix functions, reproducing kernel Hilbert spaces and interpolation, 1989
70 **Richard F. Gundy,** Some topics in probability and analysis, 1989
69 **Frank D. Grosshans, Gian-Carlo Rota, and Joel A. Stein,** Invariant theory and superalgebras, 1987
68 **J. William Helton, Joseph A. Ball, Charles R. Johnson, and John N. Palmer,** Operator theory, analytic functions, matrices, and electrical engineering, 1987
67 **Harald Upmeier,** Jordan algebras in analysis, operator theory, and quantum mechanics, 1987
66 **G. Andrews,** q-Series: Their development and application in analysis, number theory, combinatorics, physics and computer algebra, 1986
65 **Paul H. Rabinowitz,** Minimax methods in critical point theory with applications to differential equations, 1986
64 **Donald S. Passman,** Group rings, crossed products and Galois theory, 1986
63 **Walter Rudin,** New constructions of functions holomorphic in the unit ball of C^n, 1986
62 **Béla Bollobás,** Extremal graph theory with emphasis on probabilistic methods, 1986
61 **Mogens Flensted-Jensen,** Analysis on non-Riemannian symmetric spaces, 1986
60 **Gilles Pisier,** Factorization of linear operators and geometry of Banach spaces, 1986
59 **Roger Howe and Allen Moy,** Harish-Chandra homomorphisms for $\mathfrak{p}$-adic groups, 1985
58 **H. Blaine Lawson, Jr.,** The theory of gauge fields in four dimensions, 1985
57 **Jerry L. Kazdan,** Prescribing the curvature of a Riemannian manifold, 1985
56 **Hari Bercovici, Ciprian Foiaş, and Carl Pearcy,** Dual algebras with applications to invariant subspaces and dilation theory, 1985
55 **William Arveson,** Ten lectures on operator algebras, 1984
54 **William Fulton,** Introduction to intersection theory in algebraic geometry, 1984

Other Titles in This Series

53 **Wilhelm Klingenberg,** Closed geodesics on Riemannian manifolds, 1983
52 **Tsit-Yuen Lam,** Orderings, valuations and quadratic forms, 1983
51 **Masamichi Takesaki,** Structure of factors and automorphism groups, 1983
50 **James Eells and Luc Lemaire,** Selected topics in harmonic maps, 1983
49 **John M. Franks,** Homology and dynamical systems, 1982
48 **W. Stephen Wilson,** Brown-Peterson homology: an introduction and sampler, 1982
47 **Jack K. Hale,** Topics in dynamic bifurcation theory, 1981
46 **Edward G. Effros,** Dimensions and C^*-algebras, 1981
45 **Ronald L. Graham,** Rudiments of Ramsey theory, 1981
44 **Phillip A. Griffiths,** An introduction to the theory of special divisors on algebraic curves, 1980
43 **William Jaco,** Lectures on three-manifold topology, 1980
42 **Jean Dieudonné,** Special functions and linear representations of Lie groups, 1980
41 **D. J. Newman,** Approximation with rational functions, 1979
40 **Jean Mawhin,** Topological degree methods in nonlinear boundary value problems, 1979
39 **George Lusztig,** Representations of finite Chevalley groups, 1978
38 **Charles Conley,** Isolated invariant sets and the Morse index, 1978
37 **Masayoshi Nagata,** Polynomial rings and affine spaces, 1978
36 **Carl M. Pearcy,** Some recent developments in operator theory, 1978
35 **R. Bowen,** On Axiom A diffeomorphisms, 1978
34 **L. Auslander,** Lecture notes on nil-theta functions, 1977
33 **G. Glauberman,** Factorizations in local subgroups of finite groups, 1977
32 **W. M. Schmidt,** Small fractional parts of polynomials, 1977
31 **R. R. Coifman and G. Weiss,** Transference methods in analysis, 1977
30 **A. Pełczyński,** Banach spaces of analytic functions and absolutely summing operators, 1977
29 **A. Weinstein,** Lectures on symplectic manifolds, 1977
28 **T. A. Chapman,** Lectures on Hilbert cube manifolds, 1976
27 **H. Blaine Lawson, Jr.,** The quantitative theory of foliations, 1977
26 **I. Reiner,** Class groups and Picard groups of group rings and orders, 1976
25 **K. W. Gruenberg,** Relation modules of finite groups, 1976
24 **M. Hochster,** Topics in the homological theory of modules over commutative rings, 1975
23 **M. E. Rudin,** Lectures on set theoretic topology, 1975
22 **O. T. O'Meara,** Lectures on linear groups. 1974
21 **W. Stoll,** Holomorphic functions of finite order in several complex variables, 1974
20 **H. Bass,** Introduction to some methods of algebraic K-theory, 1974
19 **B. Sz.-Nagy,** Unitary dilations of Hilbert space operators and related topics, 1974
18 **A. Friedman,** Differential games. 1974
17 **L. Nirenberg,** Lectures on linear partial differential equations, 1973
16 **J. L. Taylor,** Measure algebras. 1973
15 **R. G. Douglas,** Banach algebra techniques in the theory of Toeplitz operators, 1973
14 **S. Helgason,** Analysis on Lie groups and homogeneous spaces, 1972
13 **M. Rabin,** Automata on infinite objects and Church's problem, 1972

(See the AMS catalog for earlier titles)